897

WHO EXPERT COMMITTEE ON BIOLOGICAL STANDARDIZATION

Forty-ninth Report

World Health Organization
Geneva 2000

WHO Library Cataloguing-in-Publication Data

WHO Expert Committee on Biological Standardization (1998 : Geneva, Switzerland)
WHO Expert Committee on Biological Standardization : forty-ninth report

(WHO technical report series ; 897)

1.Biological products – standards 2.Guidelines I.Title II.Series

ISBN 92 4 120897 X (NLM Classification: QW 800)
ISSN 0512-3054

Printed in Switzerland

2000/13266 – Stratcom/Schuler — 6500

Contents

WHO Expert Committee on Biological Standardization

Geneva, 19–23 October 1998

Members

Dr D. Calam, European Coordinator, National Institute for Biological Standards and Control, WHO International Laboratory for Biological Standards, Potters Bar, Herts., England

Dr M. de los Angeles Cortés Castillo, Subdirector, Quality Control, National Institute of Hygiene, Mexico City, Mexico

Dr R. Dobbelaer, Head, Biological Standardization, Louis Pasteur Scientific Institute of Public Health, Brussels, Belgium

Dr V. Grachev, Deputy Director, Institute of Poliomyelitis and Viral Encephalitides, Moscow, Russian Federation

Dr J.G. Kreeftenberg, Bureau for International Cooperation, National Institute of Public Health and Environmental Protection, Bilthoven, Netherlands (*Vice-Chairman*)

Dr F.A. Ofosu, Department of Pathology, McMaster University, Hamilton, Ontario, Canada

Dr Zhou Hai-jun, Director, National Institute for the Control of Pharmaceutical and Biological Products, Temple of Heaven, Beijing, China

Dr K. Zoon, Director, Center for Biologics Evaluation and Research, Food and Drug Administration, Bethesda, MD, USA (*Chairwoman*)

Representatives of other organizations

Council of Europe
Mr J.-M. Spieser, Head, Biological Standardization, European Department for the Quality of Medicines, Council of Europe, Strasbourg, France

European Association of the Plasma Products Industry
Dr I. von Hoegen, Director, Regulatory Affairs, Brussels, Belgium

International Association of Biological Standardization
Professor F. Horaud, Pasteur Institute, Paris, France

International Federation of Pharmaceutical Manufacturers Associations
Dr M. Duchêne, Director, Quality Control and Regulatory Affairs, SmithKline Beecham Biologicals, Rixensart, Belgium

Dr J.-C. Vincent-Falquet, Director, Regulatory Affairs, Pasteur Mérieux Connaught, Marcy l'Etoile, France

International Society of Blood Transfusion
Dr M.L. Scott, International Blood Group Reference Laboratory, Bristol, England

International Society on Thrombosis and Haemostasis
Dr D. Thomas, Kirtlington, Oxford, England

Secretariat

Dr Y. Arakawa, Director, Department of Bacterial and Blood Products, National Institute of Infectious Diseases, Tokyo, Japan (*Temporary Adviser*)

Dr D. Armstrong, Executive Director, Natal Bioproducts Institute, Pinetown, South Africa (*Temporary Adviser*)

Dr T. Barrowcliffe, National Institute for Biological Standards and Control, WHO International Laboratory for Biological Standards, Potters Bar, Herts., England (*Temporary Adviser*)

Dr E. Griffiths, Chief, Biologicals, World Health Organization, Geneva, Switzerland

Dr M.C. Hardegree, Director, Office of Vaccine Research and Review, Center for Biologics Evaluation and Research, Food and Drug Administration, Bethesda, MD, USA (*Temporary Adviser*)

Dr A.M. Padilla, Scientist, Biologicals, World Health Organization, Geneva, Switzerland

Dr F. Reigel, Director, Division of Biologicals, Swiss Federal Office of Public Health, Berne, Switzerland

Dr H. Roberts, University of North Carolina Medical School, Chapel Hill, NC, USA (*Temporary Adviser*)

Dr G. Schild, Director, National Institute for Biological Standards and Control, WHO International Laboratory for Biological Standards, Potters Bar, Herts., England (*Temporary Adviser*)

Dr W.G. van Aken, Medical Director, Central Laboratory of the Netherlands Red Cross Blood Transfusion Service, WHO International Laboratory for Biological Standards, Amsterdam, Netherlands (*Temporary Adviser*)

Dr D. Wood, National Institute for Biological Standards and Control, WHO International Laboratory for Biological Standards, Potters Bar, Herts., England (*Rapporteur*)

Introduction

The WHO Expert Committee on Biological Standardization met in Geneva from 19 to 23 October 1998. The meeting was opened on behalf of the Director-General by Dr M. Scholtz, Executive Director, Health Technology and Pharmaceuticals.

Dr Scholtz welcomed the members of the Committee, Temporary Advisers and representatives from nongovernmental organizations and industry. He emphasized the importance of the work of the Committee for both developed and developing countries. Biological medicines make an enormous contribution to public health. However, the very nature of biological products raises particular questions regarding their quality control, and he stressed that the considerable potential hazards associated with some of these substances required continuous vigilance.

Dr Scholtz recalled that in May 1997 the World Health Assembly unanimously adopted a resolution on the quality of biological products moving in international commerce (WHA50.20). The resolution recognized the special technical expertise needed for evaluating and controlling biological products, as well as the long-standing and valuable role of WHO's Biologicals unit and the Expert Committee on Biological Standardization. However, the resolution also recognized that WHO's biological standardization activities needed to be strengthened to meet the challenges of the rapid expansion and increasing complexities of the biologicals field.

In pursuance of the resolution, an independent review of WHO's activities in biological standardization was commissioned with a view to recommending action that would enable WHO to respond to scientific developments in a timely manner and to strengthen the mechanisms for providing clear norms and active leadership in promoting the quality, safety and efficacy of biological and biotechnological products. This review was now complete and a draft report of the external review team, which had consulted widely with experts inside and outside WHO, was to be discussed during the meeting.

Dr Scholtz concluded by thanking scientists who participate in WHO meetings, such as those of the Expert Committee on Biological Standardization, and in collaborative studies organized by WHO International Laboratories for Biological Standards and WHO Collaborating Centres, and those members of the Expert Panels on Biological Standardization and on Human Blood Products and Related Substances who comment on proposed standards and requirements. Dr Scholtz also thanked the various nongovernmental organizations that provide support in many ways, as well as the donors of candidate reference materials, which are, in most cases, the manufacturers of biological materials. All had contributed much to the success of WHO's international biological standardization activities.

General

Developments in biological standardization

The Committee was informed that the change of custodianship of many international reference materials, which had been made necessary by changes in the functioning of two former custodian laboratories, had now been completed. The transfer of international reference materials from the Statens Seruminstitut, Copenhagen, and the Central Veterinary Laboratory, Weybridge, England, to the National Institute for Biological Standards and Control, Potters Bar, England, had been accomplished in a safe and timely manner. This consolidation of the stocks of international reference materials had provided an opportunity to review the continued need for certain preparations and to prioritize needed replacements.

The Committee warmly welcomed the establishment of a new WHO Collaborating Centre for Biological Standardization at the Center for Biologics Evaluation and Research, Food and Drug Administration, Rockville, MD, USA. This significant development was further enhanced by a recent bilateral agreement between the Center for Biologics Evaluation and Research and the National Institute for Biological Standards and Control, to collaborate on regulatory research. The Committee strongly endorsed the development of coordinated international collaborative research initiatives in regulatory areas that concern the safety and efficacy of biological medicines of global public health importance.

The Committee commended the Secretariat for its resourcefulness in obtaining external financial support for many activities in the area of biological standardization. In addition to support provided by the WHO International Laboratories and Collaborating Centres, the Children's Vaccine Initiative had provided support for harmonization activities, the International Society on Thrombosis and Haemostasis had contributed much to the work on standards for haematological products, and the International Association of Biological Standardization had organized specialist meetings to discuss scientific issues relevant to biological standards and the control of biological products during the previous year. The Committee expressed its thanks to these organizations for their support at a time when resources for biological standardization were very limited.

The speed of dissemination of the Committee's decisions continued to be a matter of concern. The Committee had made a number of proposals to speed up the process. Rapid publication in 1997 of a summary of the major decisions stemming from the forty-seventh meeting of the Committee, together with the changes to the list of international reference materials, had proved very useful, and the Committee requested that the procedure be continued. Publication by means of the Internet was also recommended as a particularly effective way of speeding up access to new or revised documents. The wide distribution of draft documents was also endorsed, provided that the draft status of such documents was unambiguous. Distribution of drafts would facilitate adoption of a final text by national control authorities and manufacturers. Finally, the Committee considered ways to ensure that, once published, the information reached the appropriate scientific audience. A proposal from the Secretariat to republish key documents in the general scientific press was accepted.

Global project on quality assurance of plasma-derived medicinal products and plasma-fractionation activities

The Committee was informed of a global project on the quality assurance of plasma-derived medicinal products and plasma-fractionation activities which had been developed in response to the urgent need expressed by some Member States for guidance about the safety and quality of blood products. The project aimed to develop guidelines for good plasma-fractionation practices, to offer seminars and workshops on manufacturing procedures for plasma-derived products — with an emphasis on virus inactivation procedures and their validation — and to keep Member States updated on current regulatory issues in the field. The Committee strongly supported this activity. The Committee also considered that, in the future, areas such as cell and tissue banking may benefit from the work of the project.

Independent review of WHO's remit and activities in the field of biologicals

Following the adoption of a resolution at the Fiftieth World Health Assembly in May 1997 on the quality of biological products moving in international commerce (WHA50.20), an independent review was initiated to study the remit and activities of WHO in the biologicals field and in particular the work of the WHO Biologicals unit. A review team and consultative group were convened for this purpose and, after wide consultation, prepared a report that identified a number of key issues. A draft of this report was considered by the Committee. The review concluded that WHO had consistently set global standards for biological medicines and should continue to do so in the future. WHO should provide international leadership in developing, establishing and promoting authoritative documents to ensure the safety and efficacy of all biological medicines (vaccines, blood and blood products, and biological therapeutics) and the primary reference materials that enable biological standardization around the world. The draft report concluded that it was important for WHO to speak with a clear, authoritative and single "voice" on these matters.

The format of documents concerning the manufacture and control of biological medicines could be improved to accommodate the needs both for detailed technical guidance in certain countries and for a more flexible, "in principle" approach in others. The consultative process for developing these documents should be more transparent, and access to draft and final versions should be improved through greater use of modern information technology. The Committee decided that the requirements for biological substances published by WHO should be renamed "recommendations" to better reflect their nature.[1] A clear explanation of how such recommendations should be used should be included on the front page of each document.

The review recommended that the Expert Committee on Biological Standardization should be restructured as an executive and policy-making committee supported by three subcommittees of experts: one each on vaccines, blood and blood products, and biological therapeutics. The Committee recognized many advantages to this proposal, which would increase openness, efficiency, credibility and productivity. However, the Committee was strongly of the view that responsibility for overseeing the work of each of the proposed subcommittees must reside with the Expert Committee on Biological Standardization, because many issues in the control and standardization of biological medicines cut across all three areas of activity of the proposed subcommittees. It was therefore essential for a single body to ensure that uniform advice was provided in each of the areas. Similarly, it was essential for the Biologicals unit in WHO to act as the primary focus for activities and policy in the three areas of biological medicines.

Another important conclusion from the review was that WHO should retain responsibilities for reference preparations or documents concerning veterinary agents capable of infecting humans (zoonotic agents), or

---

[1] As a consequence of this decision, new documents will be issued as recommendations and a box containing an explanation of how they should be used will be included on the front page. Existing requirements will be renamed recommendations as and when they are revised. The use of the Requirements for Biological Substances numbering system will be discontinued. For further information on published Recommendations and guidelines for biological substances used in medicine and other documents, see Annex 3, p. 68 of this report. Elsewhere in this report the term "recommendations" is used, except in reference to specific existing requirements.

agents potentially threatening human health. However, WHO should not be responsible for reference preparations or documents concerning agents that infected only animals and did not pose any threat to human health. The Committee agreed with this proposal, with the comment that WHO had a duty to ensure that responsibility for useful documents or reference preparations falling into the latter category was transferred to an international organization with the relevant remit and authority in this field.

The review encouraged WHO to establish a collaborating centre for biological standardization in a laboratory in a developing country with a good record of regulatory research. This would increase the global capacity for scientific work in the area of biological standardization and control and would provide important training and staff development opportunities through interchange of staff with the existing WHO Collaborating Centres and International Laboratories. The Committee endorsed this proposal.

Reverse transcriptase associated with avian cells

The Committee was informed of progress in this area since its last meeting. The detection of low levels of reverse transcriptase in vaccines derived from chicken cells had given rise to concerns that a previously unidentified avian retrovirus could be present in vaccines prepared using chicken cells. A meeting was convened in April 1998 to review all existing scientific data, and the report of the meeting was published.[1] In summary, the low levels of reverse transcriptase activity have been shown to be associated with particles of defective endogenous avian virus and avian leukosis virus. Sequence studies have revealed potential reading frames for the *gag* and *pol* genes but not for the *env* gene (which is essential for infectivity). Extensive studies in three independent laboratories have failed to demonstrate transmission of reverse transcriptase activity or productive infection. Epidemiological studies have revealed no association between the use of chicken-cell-derived vaccines and an increased rate of detection of childhood cancers. Limited serological studies have failed to show antibodies to avian retroviral antigens after several doses of measles vaccines. The meeting concluded that vaccines prepared using chicken cells should continue to be used, provided that they met recommendations published by WHO, since the risk of vaccine-preventable disease was real and quantifiable whereas the risk posed by the chicken-cell vaccines was theoretical and remote. The meeting also concluded that WHO should establish an international task force, including scientists from academia, regulatory authorities and

---

[1] *Weekly Epidemiological Record*, 1998, 73:209–212.

industry, to coordinate collaborative research relevant to the characterization, quality control and safety assessment of all cell substrates intended for vaccine production. The Committee agreed with the conclusions and strongly endorsed the proposal to establish the task force. The Committee stressed that the task force should be concerned with all cell substrates and, as a matter of priority, plan and develop sensitive and standardized assays for reverse transcriptase activity in vaccines grown on non-avian cell substrates. Furthermore, the Committee recommended that a workshop be organized on the acceptability of cell substrates for use in biological medicines.

Comments from users of recommendations (requirements) and guidelines published by WHO and international reference materials

The independent review of WHO's activities in the area of biologicals identified a need for improved consultation with, and feedback from, users of documents published by WHO and of international reference materials. The Committee was therefore very pleased to receive a report on users' experiences from WHO's Global Programme on Vaccines and Immunization.

The potency tests for diphtheria and tetanus toxoids described in the 1990 Requirements for Diphtheria, Tetanus, Pertussis and Combined Vaccines (WHO Technical Report Series, No. 800) have caused difficulties for several manufacturers and national control authorities. Many laboratories have been unable to obtain the required supply of animals for the potency test, and have been using methods other than those given in the recommendations (requirements) published by WHO. Validation of reference materials has also proved to be a problem; and ambiguities in wording have caused difficulties in the interpretation of certain sections (e.g. the mouse weight-gain test for pertussis toxicity). The question of whether there is a continuing need to perform the abnormal toxicity test on diphtheria, tetanus and pertussis vaccine has also been raised, particularly by manufacturers.

The 1994 Requirements for Measles, Mumps and Rubella Vaccines and Combined Vaccine (Live) (WHO Technical Report Series, No. 840) have also been found to contain ambiguities. In particular, there were inconsistencies with respect to the test for neurovirulence, and it is unclear whether it should be performed on master or working seed lots. The Committee was also informed of concerns of the National Institute of Biological Standards and Control, Potters Bar, regarding the value of the neurovirulence test for certain vaccines. The Committee therefore recommended a review of the need for, and standardization of, neurovirulence tests (except those for oral poliomyelitis vaccine and yellow fever vaccine), where such tests are specified in recommendations (requirements) published by WHO.

Another area of concern is the use of the International Reference Reagent for Plasma-derived Hepatitis B Vaccine for Immunogenicity Studies. Although the reference material has not been validated for in vitro tests, evidence from visits to manufacturers and national control authorities indicates that it has sometimes been used as calibrant for in vitro assays. Furthermore, a collaborative study organized by the European Department for the Quality of Medicines with vaccines from three manufacturers showed that within- and between-laboratory agreement was improved with product-specific reference preparations. As a wide range of recombinant vaccines are currently manufactured around the world, further studies are required to determine whether this conclusion is more widely applicable.

The Committee therefore requested that the National Institute for Biological Standards and Control, Potters Bar, conduct a review of reference materials required for standardization of hepatitis B vaccines.

International nonproprietary names

The Committee discussed the International Nonproprietary Names (INN), which are unique and globally agreed names for pharmaceutical substances. The Committee was informed that there have recently been increasing numbers of requests to name products manufactured by biotechnological processes and there have been calls for the assistance of the Expert Committee in this area.

The Expert Committee agreed in principle to advise on proposed INN but concluded that further information was necessary on the possible roles, responsibilities and expectations for it in this area. Issues with wide-ranging implications, such as proposals to establish an INN for a new class of product, would need discussion in the Expert Committee. The Committee requested the Secretariat to explore the matter further and to present a report at its next meeting.

International guidelines, recommendations and other matters related to the manufacture and quality control of biologicals

Recommendations for *Haemophilus influenzae* type b conjugate vaccine

The Committee noted a draft of proposed updated 1990 requirements for *Haemophilus influenzae* type b conjugate vaccine (BS/98.1874). The document had been prepared following extensive discussions by experts and a WHO/National Institute for Biological Standards and Control, Potters Bar, informal workshop in 1998 on physicochemical procedures appropriate for the characterization of *Haemophilus influenzae* type b conjugate vaccines.

The Committee agreed that this document, which, in accordance with its earlier decision, would be renamed Recommendations for *Haemophilus influenzae* type b Conjugate Vaccine, should be used as a pilot project for the new consultative process recommended by the independent review. Thus, after revision by the Committee to take account of comments received in the usual way, the document would be circulated in the scientific community to provide an opportunity for further public comment. The period for public comment would be short (about 3 months). Subject to confirmation by the consultative exercise, the Committee gave approval for the document to be adopted and appended to the report in the usual way (Annex 1).

Recommendations for poliomyelitis vaccine (oral)

The Committee was informed of progress in the development of alternative assays for neurovirulence in oral poliomyelitis vaccine. Since the Committee had last met, a regulatory decision-making procedure for the transgenic mouse model had been developed and validated in a collaborative study organized by WHO. A database was now being assembled, using these decision-making procedures, on vaccines previously tested in monkeys. The Committee commended the investigators for their remarkable progress to date. The Committee suggested certain additional data they would like to see in support of an anticipated proposal to change the 1990 Requirements for Poliomyelitis Vaccine, Oral to allow a transgenic mouse test as an alternative to the current monkey test. These data should be available for consideration at the next Expert Committee meeting. To ensure maximum flexibility in supply of the transgenic mouse in the future, the Committee recommended the transfer of breeding nuclei of the mouse line, TgPVR21, from the single laboratory that has so far very generously supplied all mice for the study to other collaborating organizations. Transfer should commence as soon as possible since there would inevitably be a certain period necessary to establish mouse populations at the new facilities. The needs of poliomyelitis vaccine manufacturers in developing countries should be taken into account in decisions concerning the location of animal supply facilities. The Committee also recommended that laboratories wishing to use the transgenic mouse model should comply with the procedures and standards published by

WHO for the housing of animals.[1] The Committee recommended that the Secretariat investigate, and report on, how compliance with the proposed procedures and standards for testing neurovirulence could be monitored in the future.

The Committee was informed of progress with the MAPREC (mutant analysis by polymerase chain reaction and restriction enzyme cleavage) assay. As recommended at the forty-eighth meeting of the Committee, the Secretariat had initiated an enquiry to obtain MAPREC data for poliovirus type 3 from as wide a spectrum of manufacturers of poliomyelitis vaccines as possible. The data would help guide decisions on how to use the MAPREC assay for regulatory purposes. The outcome of a review of these data would be reflected in a full revision of the 1990 Requirements, planned for submission to the next Expert Committee meeting. Collaborative studies to evaluate candidate standard reagents for MAPREC assay of poliovirus types 1 and 2 were also under way.

The Committee reviewed a proposed addendum to the 1990 Requirements (BS/98.1877). The requirements had last been revised by the Committee in 1989 and additions were now required in four areas. These were: a recommendation for the working virus seed to be free of detectable sequences of simian virus 40 (SV 40); guidance on technical performance of the MAPREC assay for poliovirus type 3; the need for increased levels of laboratory containment of wild polioviruses used as controls in the rct40 assay; and guidance on antibody screening tests for animals from closed primate colonies used as sources of primary monkey kidney cells. After making some modifications, the Committee adopted the text as an addendum to the 1990 Requirements for Poliomyelitis Vaccine (Oral), Addendum 1998 and agreed that it should be annexed to its report (Annex 2).

Guidelines for the production and control of the acellular pertussis component of monovalent or combined vaccines

The Committee was informed of a meeting of a WHO Working Group on the Standardization and Control of Pertussis Vaccines. The Working Group had provided feedback to the Committee on the recently adopted Guidelines for the Production and Control of the Acellular Pertussis Component of Monovalent or Combined Vaccines (WHO Technical Report Series, No. 878).

[1] Maintenance and distribution of transgenic mice susceptible to human viruses: memorandum from a WHO meeting. *Bulletin of the World Health Organization*, 1993, **71**:497–502.

The guidelines were concerned with vaccines shown to be safe and effective in scientifically well-controlled clinical studies. Difficulties had been identified because this criterion would potentially exclude from the international market certain vaccines in routine use in Japan. Concerns about whole-cell pertussis vaccines had led to acellular pertussis vaccines being introduced into Japan's national vaccination programme without the benefit of a full double-blind clinical efficacy trial. The Committee emphasized that it was not the intention of the guidelines to seem to exclude effective vaccines of this kind and agreed that in such cases the results of product-specific post-marketing surveillance and epidemiological data could be used to show efficacy.

The Committee also confirmed acceptance for potency estimation of the modified intracerebral challenge assays as an alternative to the immunogenicity test. It was agreed that the modified assay could be used to demonstrate immunogenic potential, but would not be considered to reflect clinical efficacy. However, considerable confusion could occur if the potency of acellular pertussis vaccines as estimated in the modified intracerebral challenge test were to be expressed in the International Units that had been assigned to the International Standard for whole-cell pertussis vaccine. Although these units are accepted as indicative of the protection induced by whole-cell pertussis vaccines in children, the Committee recommended that they should not be used to indicate the protective immunity induced by acellular pertussis vaccines.

The Committee was also informed of an aerosol challenge model that, in some laboratories, had been shown to reflect the clinical efficacy of acellular pertussis vaccines. The Committee recommended that validation of both the aerosol and the modified intracerebral challenge assays should be pursued by the Working Group.

The Working Group had also considered the mouse toxicity test performed in Japan as a test for lot release. The test is based on rectal temperature measurement after histamine challenge, and appears to be very sensitive. It was recommended that work be undertaken to compare the test described in the guidelines with that measuring rectal temperature modification.

The Committee was informed that the Working Group had discussed the hypothesis that whole-cell pertussis vaccines in current use might not protect against emergent new variant strains of *Bordetella pertussis* and had concluded that the available data did not support the hypothesis.

List of recommendations for biological substances and other sets of recommendations: proposed new format, electronic publication and proposed discontinuations

A list of recommendations (formerly requirements) for biological substances and other sets of recommendations has been published as an annex to each Expert Committee report. The recommendations were listed sequentially in chronological order of establishment.[1] To facilitate ease of reference, the Secretariat has recently prepared a revised annex in which recommendations were listed alphabetically, and in which only the current version of each was listed. A number of recommendations and other documents thought to be no longer relevant had been identified, and their discontinuation was proposed. Finally, to improve access to the annex document, it was proposed that the list be published by means of the Internet as well as in the conventional manner in the WHO Technical Report Series.

The Committee welcomed the revised format for the list, which should improve transparency, and also encouraged publication by means of the Internet, which should improve access. The Committee agreed with the principle that recommendations that were no longer relevant should be discontinued. However, it was considered that wider consultation was needed to decide whether specific recommendations should be discontinued. Therefore, the scientific community should be informed of the recommendations that were candidates for discontinuation, and decisions should be made in the light of responses received at the next ECBS meeting. The proposed discontinuations were the 1966 Requirements for Smallpox Vaccine (WHO Technical Report Series, No. 323) and the 1968 Requirements for Cholera Vaccine (WHO Technical Report Series, No. 413).

However, the Committee agreed that a Report of a WHO Working Group on the Standardization of Human Blood Products and Related Substances (WHO Technical Report Series, No. 610) could be discontinued immediately, since many parts of the document had already been superseded.

The Committee also requested that the reports of all consultative groups in the field of biologicals convened by WHO within the past 5 years be listed. These reports constituted a valuable scientific resource that would be of benefit to the wider scientific community. Once established, such a list could be updated on an annual basis.

The Committee agreed that the revised list entitled "Recommendations and guidelines for biological substances used in medicine and other documents", should be attached to its report (Annex 3).

[1] See, for example, WHO Technical Report Series, No. 878, pp. 95–101.

Work in progress and proposed new work on recommendations and guidelines

The Secretariat reported that revisions of a number of existing recommendations (formerly requirements) were in progress. The 1991 Requirements for Influenza Vaccines (Inactivated) needed to provide detailed guidance for manufacturers switching from production in eggs to production in cell substrates. An informal workshop on this topic had already been held at the National Institute for Biological Standards and Control, Potters Bar, and a small drafting group had been requested to prepare draft revised requirements. A consultation would subsequently review and prepare a final draft for consideration by the Committee.

A first draft of revised Guidelines for the Preparation, Characterization and Establishment of International and Other Standards and Reference Reagents for Biological Substances had already been prepared. The Committee recommended that guidance on calibration of national reference reagents be included in a second draft. The Committee was pleased to learn of the intention to submit the revised document for consideration at its next meeting.

Major changes to the 1990 Requirements for Diphtheria, Tetanus, Pertussis and Combined Vaccines (WHO Technical Report Series, No. 800) were foreseen. In particular, it would be timely to review available data to determine whether a change in current potency requirements could be made. In addition, certain standardization problems with existing assays should be addressed. A scientific meeting was required to identify further studies needed on potency assays, and an informal consultation would ultimately be required to consider any proposed changes to the requirements.

The 1971 Requirements for Snake Antivenins (WHO Technical Report Series, No. 463) were considered in urgent need of revision. Manufacturing processes had progressed significantly since the requirements were established, and new issues had been identified concerning potential adventitious agents. The Secretariat was requested to assemble a small team to prepare a first draft of a revision and subsequently to convene a workshop to review the document.

The need had been identified for an addendum to the 1996 Requirements for the Use of Animal Cells as in vitro Substrates for the Production of Biologicals (WHO Technical Report Series, No. 878) to cover the acceptability of primary cell substrates for the production of live virus vaccines, as well as the viral safety of animal sera and other biological substances used during propagation of cells, and the implications of changes in cell growth media. The Committee supported the preparation of such an addendum and also requested the Secretariat to convene workshops to consider these and other issues raised by the introduction of new cell lines and new production methods such as transgenic plants.

The 1990 Requirements for Poliomyelitis Vaccine, Oral (WHO Technical Report Series, No. 800) were also in need of revision to accommodate advances in quality control techniques such as the MAPREC assay and the use of transgenic mice for neurovirulence tests. Although an addendum to the existing requirements was considered and adopted (Annex 2), the Committee agreed with the need for a subsequent full revision. A consultation would be required to consider the draft revision before submission to the Committee.

The Committee was informed that a number of new documents were at various stages of development. The priorities identified by the Committee were recommendations for the design and interpretation of stability studies of biological substances used in medicine, especially vaccines; recommendations for rotavirus vaccines; recommendations for live Japanese encephalitis vaccine; recommendations for viral inactivation and/or removal procedures and their validation for plasma and plasma-derived medicinal products; and recommendations for live tetravalent dengue vaccine. The Committee also identified the need for recommendations for combined vaccines, new cholera vaccines and pneumococcal conjugate vaccine and requested the Secretariat to continue development of the relevant documents as resources would permit.

International reference materials

Review of existing reference materials: proposed database and electronic publication

The Committee was informed that the Secretariat had initiated a major review of existing biological reference materials with the goal of updating the catalogue, last published in 1991.[1] An additional goal was to make the catalogue available by means of the Internet. A comprehensive electronic database had been prepared that contained key information, including references to the relevant background documents containing scientific information about the reference materials. The Committee welcomed this development and agreed that the database should be updated to include reference materials established, and exclude reference materials discontinued, at this meeting and agreed that the list of current reference materials should be annexed to its report (Annex 4). The Secretariat was also encouraged to publish the database by means of the

[1] *Biological substances: International Standards and Reference Reagents, 1990.* Geneva, World Health Organization, 1991.

Internet and, resources permitting, to incorporate links to the original background documents.

The Committee noted that the review had revealed the need to monitor stocks of international reference materials and their rates of issue, and recommended that all custodian laboratories should provide such information annually to the Secretariat. This would enable the Biologicals unit to manage its work on international reference materials more efficiently.

Proposals for replacement or discontinuation

The Committee considered a review of the status, and proposals for replacement and discontinuation, of international reference materials held and distributed by the National Institute for Biological Standards and Control, Potters Bar (BS/98.1892, BS/98.1892 Add. 1), by the Central Laboratory of the Netherlands Red Cross Blood Transfusion Service, Amsterdam (BS/98.1893), and by the Centers for Disease Control and Prevention, Atlanta (BS/98.1893). The Committee confirmed the procedure adopted at its forty-eighth meeting according to which certain proposals to discontinue international reference materials were published for comment before a final decision was taken. On the basis of the above-mentioned documents and further information received during the meeting, the Committee agreed to discontinue the international reference materials for the following substances held at the National Institute for Biological Standards and Control, Potters Bar: demeclocycline, doxycycline, minocycline, oxytetracycline, tetracycline, hyaluronidase and hygromycin B.

A nominal archival stock (usually 50 ampoules) of all seven discontinued materials would be retained indefinitely at the National Institute for Biological Standards and Control, Potters Bar. The Committee agreed that remaining stocks of the first five preparations, all antibiotics, should be offered to the WHO Collaborating Centre for Chemical Reference Substances, Stockholm. The Committee noted that the International Pharmaceutical Federation's (FIP) International Centre for Pharmaceutical Enzymes holds a recently prepared reference material of hyaluronidase calibrated by the current assay method against the discontinued International Standard, and that this preparation would become the primary international reference material.

The Committee was informed that several reference materials held at the Centers for Disease Control and Prevention, Atlanta, were no longer available for distribution and therefore agreed to discontinue the International Reference Reagents for: histoplasmin for H and M immunodiffusion test, histoplasmin antiserum, rabbit, for H and M immunodiffusion test, *Mycoplasma pneumoniae* antiserum, equine, and parainfluenza virus antisera, equine.

The Committee recommended discontinuation of the following international reference materials in 1999, provided that no objections were raised:

- the International Standard for Anti-A,B Blood-Typing Serum
- the International Reference Preparation of Nisin
- the International Reference Preparation of Candicidin
- the International Standard for Rolitetracycline
- the International Standard for Desmopressin
- the International Reference Preparation of Gonadorelin, for Bioassay
- the International Reference Preparation of Parathyroid Hormone, Bovine, for Bioassay
- the International Reference Reagents for subtype-specific antisera to hepatitis B surface antigens, i.e. anti-HBs/ay serum (goat), anti-HBs/ad serum (goat), anti-HBs/ay serum (guinea-pig), anti-HBs/ad serum (guinea-pig) and anti-HBs/ar serum (rabbit)
- the International Standard for FITC-Conjugated Sheep Anti-Human Ig
- the International Standard for FITC-Conjugated Sheep Anti-Human IgG (Anti-γ-Chain)
- the International Standard for FITC-Conjugated Sheep Anti-Human IgM (Anti-μ-Chain).

The Committee requested the National Institute for Biological Standards and Control, Potters Bar, to proceed with obtaining and evaluating candidate preparations for replacement of the following materials for which stocks had become depleted:

- the International Standard for Amphotericin B
- the International Standard for Vancomycin
- the second International Standard for Blood Coagulation Factor II and X, Concentrate, Human
- the International Reference Reagent for Plasma Fibrinogen, Human
- the International Standard for Platelet Factor 4
- the second International Standard for Salmon Calcitonin
- the third International Standard for Human Chorionic Gonadotrophin
- the third International Standard for Follicle-stimulating and Luteinizing Hormone
- the International Standard for Somatropin (Recombinant DNA-derived Human Growth Hormone).

The Committee noted the need to confirm whether a replacement for the International Standard for Human Thrombin was required. The

Committee also recalled its decision at the forty-eighth meeting in 1997 that a working group should be established to evaluate the need for production of botulinum toxoids and antitoxins for therapeutic use, as well as the need for reagents for standardization. Replacement of existing standards will depend on this evaluation.

The Committee agreed that no action should be taken to replace the following reference materials for which stocks had become depleted:

- the International Reference Preparation of Spiramycin
- the International Reference Preparation of Pyrogen

  or for the following materials, to which low priority had been assigned by a consultation (see page 17):
- the International Standard for Anti-E, Complete Blood-Typing Serum
- the International Standard for Anti-C, Incomplete Blood-Typing Serum, Human
- the International Standard for Anti-C, Complete Blood-Typing Serum.

The Committee noted that an enquiry had shown continuing demand for an International Reference Preparation of Opacity and that a replacement material for the fifth International Standard for Opacity was being evaluated.

The Committee deferred taking action on preparations of veterinary reference materials pending further assessment of the recommendations of the independent review and the Committee's decisions regarding responsibilities for veterinary reference materials.

The Committee noted that stocks of the first International Reference Reagent for Apolipoprotein B were low. This material had been developed in collaboration with the International Federation of Clinical Chemistry (IFCC). The Committee recommended that the Secretariat discuss with IFCC strategies for replacement, if appropriate.

Work in progress and proposed new work: general report

The Committee was presented with a list of work in progress and of proposed new work by the National Institute for Biological Standards and Control, Potters Bar. The Committee considered that it would be useful for such lists to include an indication of the purpose of the proposed reference materials. The Committee welcomed this initiative on the part of the Institute and recommended that all the WHO International Laboratories and relevant WHO Collaborating Centres prepare such a list in the future, which would enable better coordination by the Secretariat.

WHO consultation on review and replacement of International Standards for blood-grouping reagents

The Committee noted a report of a meeting to develop a plan to review and update as necessary WHO International Standards for blood-grouping reagents. There was an urgent need for reference materials suitable as minimum potency standards for blood-grouping reagents, which would differentiate safe from unsafe reagents. The meeting considered that the reference materials should be suitable for use with slide, tube and microplate and other agglutination test methods; they should also be suitable as standards for both polyclonal and monoclonal reagents. However, the candidate reference materials need not be designed to determine specificity. The meeting had reviewed existing international standards for blood-grouping reagents and had prioritized the need for replacements, on the basis of remaining quantities, age and infectious disease marker status, as follows: first, anti-D, complete (IgM) and anti-D, incomplete (IgG); second, anti-A and anti-B; and third, anti-IgG. The WHO International Laboratories at the National Institute for Biological Standards and Control, Potters Bar, and the Central Laboratory of the Netherlands Red Cross Blood Transfusion Service, Amsterdam, and the WHO Collaborating Centre at the Center for Biologics Evaluation and Research, Bethesda, MD, together with expert panels appointed by the International Society for Blood Transfusion and the International Committee for Standardization in Haematology, as well as representatives of reagent manufacturers, would be involved in this project. The Committee appreciated the input of the meeting to help establish priorities and endorsed its proposals.

The meeting also considered that there was either no longer a need, or an uncertain need, for certain International Standards for blood-grouping reagents and that they could either be discontinued or not replaced when existing stock ran out. The standards thus identified were those for the following blood-typing sera: anti-A,B, anti-C, complete, anti-C, incomplete and anti-E, complete. The Committee recommended that these materials be added to the list of International Standards proposed for discontinuation that was to be published after its meeting.

Working group on reference preparations for testing diagnostic kits used for detection of hepatitis B and C and human immunodeficiency virus in blood screening

The Committee noted a report of a working group established to develop reference preparations for testing diagnostic kits used for detection of hepatitis B surface antigen (HBsAg) and antibodies to hepatitis C (anti-HCV) and human immunodeficiency virus (anti-HIV) in blood screening.

Three candidate reference materials had been studied and had been selected by the working group for preparation of a HBsAg reference dilution panel. A series of five dilutions, ranging from 25 to 0.04 IU/ml, would be prepared and assessed in a collaborative study.

The working group had proposed that an anti-HCV (genotype 1) candidate reference material be prepared from a pool of 12 high-titre genotype-1 donations that had been characterized by members of the group. It was intended that the candidate anti-HCV genotype-1 reference material would be further characterized by collaborative study. The working group considered it would be appropriate to establish a reference dilution panel comprising antibodies to different HCV genotypes. The working group had therefore accepted an offer from the National Institute for Biological Standards and Control, Potters Bar, of anti-HCV genotype-2 and genotype-3 donations and agreed to try to obtain anti-HCV genotype-4 and genotype-5 donations. The group considered it unlikely that an anti-HCV genotype-6 donation could be identified.

The working group had endorsed a proposal for a reference panel of anti-HIV reagents comprising donations of different subtypes: HIV-1 subtypes A, B, C, D, E and F, as well as Group O and HIV-2. Suitable donations for several components of the proposed panel had been identified, and members of the working group had agreed to attempt to identify suitable donations for the rest.

The Committee commended the working group for the excellent collaboration that had been established through this project. The Committee also considered that the working group was achieving its aim of advancing the scientific basis for the establishment of WHO reference materials for the quality control of kits for detection of HBsAg, anti-HCV and anti-HIV.

Standardization of diagnostic tests for prions

The Committee was informed of a proposed working group to develop candidate reference materials for the prion diagnostic tests that were becoming available. Reliable methods for the accurate diagnosis of bovine spongiform encephalopathy (BSE) and other human and animal transmissible spongiform encephalopathies (TSEs) are urgently needed, and the Committee considered that this was of the utmost importance and that the proposed working group was therefore a timely development.

Antibodies

Clostridium perfringens beta antitoxin, equine

The Committee noted that stocks of the International Standard for *Clostridium perfringens* beta antitoxin, a preparation of hyperimmune equine serum, were exhausted and that a replacement preparation was needed (BS/98.1892). On the basis of information contained in the background document, the Committee therefore established a preparation in ampoules, coded 2Cp beta At, as the second International Standard for *Clostridium perfringens* beta antitoxin and, on the basis of the report (BS/85.1484), assigned to it a potency of 4770 IU per ampoule.

Anti-hepatitis A immunoglobulin and serum, human

At its forty-eighth meeting the Committee had been informed that the International Reference Preparation of Hepatitis A Immunoglobulin, established in 1981, had recently been found positive for hepatitis C virus RNA by polymerase chain reaction. The Committee had therefore recommended studies of a replacement material. The Committee noted the results of a collaborative study performed in 16 laboratories in 11 countries of an immunoglobulin preparation coded 97/646, which was negative for all tested virus markers (BS/98.1878, BS/98.1878 Add. 1). The Committee agreed to establish the material as the second International Standard for Hepatitis A Immunoglobulin and assigned to it a potency of 49 IU per ampoule.

The Committee also considered a proposal in the same document to establish a preparation of human serum containing hepatitis A antibodies, coded 97/648, as an international reference material. The candidate serum standard had been included in the collaborative study to determine whether there was a need for a separate serum standard for assay of hepatitis A antibodies in serum. The Committee concluded that the study did not provide sufficient evidence to justify a separate serum standard and decided not to establish the candidate preparation as a reference material. However, the Committee requested that additional information regarding the need for a serum standard be sought and reported back to the Committee.

Antigens and related substances

Pertussis vaccine: third International Standard

The Committee noted the results of a collaborative study (BS/98.1880) performed in four laboratories in three countries testing a batch of pertussis vaccine, coded 66/303, against the second International Standard for Pertussis Vaccine. The study was carried out in response to an urgent need to replace the second International Standard, stocks of which would probably become exhausted before a planned candidate replacement, coded 94/532, could be fully evaluated. The preparation coded 66/303 was in fact a second aliquot of the same diluted suspension of *Bordetella pertussis* established as the second International Standard, and had been filled into ampoules in exactly the same manner. Results of physicochemical tests, including that for moisture content, showed that 66/303 and the second International Standard were indistinguishable. Another aliquot of the preparation coded 66/303 had previously been established as the first British Standard for pertussis vaccine, with an assigned potency of 46 IU per ampoule.[1]

The collaborative study (BS/98.1880) showed that potency previously assigned to 66/303 remained appropriate and was indistinguishable from that of the second International Standard.[1] The Committee therefore established the preparation coded 66/303 as the third International Standard for Pertussis Vaccine, and assigned to it a unitage of 46 IU per ampoule. The Committee further recommended that evaluation of the candidate preparation 94/532 should be completed as soon as possible.

Blood products and related substances

Blood coagulation factor VII concentrate

The Committee noted the results of a collaborative study on a preparation of factor VII concentrate performed by 16 laboratories in 10 countries (BS/98.1876). The candidate preparation had been assayed against both the second International Standard for Blood Coagulation Factors II, VII, IX, X, Plasma, Human, and against normal plasma pools.

The Committee further noted that the collaborative study report had been peer-reviewed and approved by the Scientific and Standardization Committee of the International Society on Thrombosis and Haemostasis. On the basis of the study, and in light of the satisfactory stability of the preparation, the Committee established the preparation, coded 97/592, as the International Standard for Blood Coagulation Factor VII Concentrate, Human, and assigned to it a potency of 6.3 IU of factor VII activity per ampoule. The Committee noted that the addendum to the background report was necessary as it provided information on some of the physicochemical tests carried out on the material. The Secretariat was requested to ensure that the addendum and report were collated and archived so that the record would be as complete as possible.

---

[1] Seagroatt V, Sheffield F. *Journal of Biological Standardization*, 1981, 9:351–365.

Blood coagulation factor VIII concentrate

The Committee noted the results of a collaborative study on a preparation of factor VIII concentrate, coded 97/616, performed by 33 laboratories (BS/98.1888). This material was a preparation of recombinant factor VIII intended as a replacement for the fifth International Standard for Blood Coagulation Factor VIII Concentrate. All previous international standards had been derived from plasma. The study demonstrated that the proposed candidate was acceptable in terms of parallelism of the dose–response relationships and had good stability. After discussion, the Committee adopted the material coded 97/616 as the sixth International Standard for Blood Coagulation Factor VIII Concentrate, Human, and assigned to it a potency of 8.5 IU of factor VIII activity per ampoule. This potency was consistent with recommendations from the Scientific and Standardization Committee of the International Society on Thrombosis and Haemostasis.

The Committee also recommended that the recommendations of the Scientific and Standardization Committee of the International Society on Thrombosis and Haemostasis be added as an addendum to BS/98.1888 so that the record would be as complete as possible.

Blood coagulation factor VIII and von Willebrand factor in plasma

The Committee reviewed the results of a collaborative study to calibrate a candidate replacement material against the third International Standard for Blood Coagulation Factor VIII and von Willebrand Factor in Plasma, in which 25 laboratories in 10 different countries participated (BS/98.1881). The material, coded 97/586, was found suitable and was established as the fourth International Standard for Factor VIII and von Willebrand Factor, Plasma, Human. The Committee assigned to it potencies of 0.57 IU per ampoule for factor VIII clotting activity, 0.89 IU per ampoule for factor VIII antigen, 0.79 IU per ampoule for von Willebrand antigen, and 0.73 IU per ampoule for von Willebrand ristocetin cofactor activity. The Committee noted that the material had a low potency for factor VIII clotting activity, as a result of the time necessary to obtain microbiological test results having precluded ampoule filling on the same day as dilution. The Secretariat was requested to obtain information on the impact of this low potency for users of the reference material. The Committee also recommended that the next replacement standard should have a higher potency for factor VIII clotting activity.

Blood coagulation factor IXa, human

The Committee was informed of a collaborative study on a preparation of activated factor IX, coded 97/562, performed by nine laboratories

(BS/98.1879 Rev.1). The Committee was also informed of concerns over an approximately threefold discrepancy between the potency proposed in the report of the collaborative study and the potency found by the manufacturer who had provided the bulk material. The reason for the discrepancy was the subject of an ongoing scientific investigation, and the Committee therefore postponed a decision on establishment of this material as a standard until further data were available.

Haemoglobincyanide

The Committee noted a report of a collaborative study on a proposed replacement reference material of haemoglobincyanide (HiCN), donated by the International Council for Standardization in Haematology, for use in the standardization of the spectrophotometric determination of haemoglobin concentration in blood using the HiCN method (BS/98.1886). The Committee noted that this material was of bovine rather than human origin. The candidate material was found suitable for its intended purpose in the collaborative study and showed adequate stability. The Committee established the preparation, coded 522, as the sixth International Standard for Haemoglobincyanide and assigned an activity of 49.79 μmol/litre. The Committee further noted that a separate aliquot of the same batch was established by the Community Bureau of Reference as Certified Reference Material (CRM) 522 in 1994.

The Committee recommended that all stocks of the standard should be transferred to the National Institute for Biological Standards and Control, Potters Bar, for relabelling, storage and distribution. The Committee requested the National Institute to initiate work as soon as possible to identify, prepare and study a candidate replacement standard, as stocks of the sixth International Standard were limited.

WHO informal consultation on standardization of unfractionated heparin and proposed fifth International Standard for Unfractionated Heparin

A group of experts was convened by the WHO Biologicals unit to consider the report of a collaborative study that aimed to replace the fourth International Standard for Heparin, an unfractionated preparation, and to consider ways to reduce or eliminate the long-standing discrepancy between the International Unit and the USP unit for unfractionated heparin. The group had concluded that the material coded 97/578 be recommended as the fifth International Standard for Unfractionated Heparin.

The Committee reviewed the collaborative study performed by 24 laboratories (BS/98.1875) and agreed to establish the preparation coded 97/578, which showed good stability, as the fifth International Standard for

Unfractionated Heparin, with an assigned potency of 2031 IU per ampoule. The Committee strongly endorsed a conclusion from the group of experts that the Secretariat, in collaboration with the International Society on Thrombosis and Haemostasis, should establish a working group to examine the feasibility of developing new assay methods with the aim of improving international harmonization in the field of heparin standardization.

Plasmin

The Committee noted the report on a preparation of plasmin (INN = fibrinolysin), coded 97/536, that was studied as a candidate replacement in a collaborative study involving six laboratories in six countries (BS/98.1887). The results of the study had previously been reviewed by the Subcommittee on Fibrinolysis of the Scientific and Standardization Committee of the International Society on Thrombosis and Haemostasis, which had concluded that the material was suitable for its intended use. The Committee concurred with this view and established the preparation coded 97/536, which was adequately stable, as the third International Standard for Plasmin and assigned to it a potency of 5.3 IU per ampoule. The Committee noted that, in stability studies of the preparation, the clot lysis assay, an assay of biological function, was more sensitive than the more widely used chromogenic assay. The Committee recommended further study of the relevance of this finding and similar discrepancies between assays for other blood products, and requested that a further report be presented at its next meeting.

The Committee agreed that comments from the Subcommittee on Fibrinolysis should be appended to document BS/98.1887 and included in the information distributed with the product.

Cytokines

WHO informal consultation on cytokine standards

As requested at the forty-eighth meeting of the Committee, a WHO informal consultation on cytokine and growth factor standards, held at the Center for Biologics Evaluation and Research, Bethesda, MD, had reviewed proposals for candidate standards to be submitted to the Committee. The consultative group had concluded that the proposal to establish a reference reagent for consensus interferon alpha be withdrawn until 2001, when results of the manufacturer's study would be available. The consultative group had also recommended that proposals to establish international reference materials for hepatocyte growth factor/scatter factor and insulin-like growth factor 2 needed considerable clarification and should also be withdrawn. The Committee concurred with these decisions and considered that pre-assessment of proposals by a group of experts, as had occurred in this case, was an extremely welcome development. The consultative group had also identified priorities for future standardization. These were to evaluate the suitability of existing reference reagents as international reference materials for interleukin-11, r nerve growth factor and stem cell factor. There was also a need to establish reference reagents for soluble tumour necrosis factor receptor FcP75 and thrombopoietin. Work was also under way to establish reference reagents for transforming growth factor beta-3 and interleukin-17. The Committee agreed with these proposals and re-emphasized that work should not be started on candidate reference materials until order of priority had been agreed.

The consultative group sought advice from the Committee on the production of reference materials for calibration of immunoassays. The Committee considered that the study of reference materials intended for diagnostic use only had a lower priority than the study of materials for which there was an actual or potential therapeutic use. The Committee recognized the need to produce guidelines for the consultative group, and that these guidelines could serve as models for similar groups established by WHO in other specialized areas. The Secretariat was requested to develop appropriate guidance for the use of such groups.

Finally, the consultative group had set up a subgroup to develop a standard format for the information required in documents to be submitted to the Committee. This development was welcomed by the Committee.

Activin A, human, recombinant

The Committee noted the report on a preparation of human recombinant activin A (BS/98.1882), and on this basis established the preparation, coded 91/626, as the first Reference Reagent for Activin A, Human, Recombinant and assigned to it a potency of 5 units per ampoule. The Committee further noted that the nominal content of activin A in each ampoule was 5 femtograms.

Interleukin-2 soluble receptor

The Committee noted a report of a collaborative study carried out in nine laboratories in six countries of a candidate reference material for interleukin-2 soluble receptor (BS98/1889). The Committee was unable to make a decision on the suitability of the candidate material, coded 97/600, because essential information was not presented in the report. The Committee requested the authors to redraft their report to provide the necessary information on the study design, the samples included in the study, the statistical method used and results obtained.

The Committee was also unsure of the real or potential need for this reference material and requested that a decision-tree be developed by the cytokine standards consultative group in line with recommendations made previously (see WHO Technical Report Series, No. 858, p. 5).

Sex hormone binding globulin

The Committee noted a report of a collaborative study performed by nine laboratories in three countries of a proposed reference material for sex hormone binding globulin (BS/98.1884). The study showed that expression of results relative to the candidate reference preparation improved between-laboratory variation and that the candidate material had adequate stability. The Committee therefore established the preparation, coded 95/560, as the International Standard for Sex Hormone Binding Globulin, Human, and assigned to it a potency of 107 IU per ampoule where, for the purpose of this preparation, 1 IU is equivalent to 1 nmol of sex hormone binding globulin per litre.

The Committee noted that data from bioassays and immunoassays had been combined to obtain the assigned potency. The Committee concurred with this approach, which was accepted practice in establishing potencies for international standards, and also considered that it was timely, because of the planned revision of the guidelines published by WHO on the preparation, characterization and establishment of international and other standards and reference reagents for biological substances, to review the procedure. The Committee therefore requested the Secretariat to arrange a meeting to discuss the scientific basis of combining data from different types of assay.

Miscellaneous

Botulinum type A toxin

The Committee noted a report of a collaborative study performed in 10 laboratories in five countries on three candidate reference materials for botulinum type A toxin (BS/98.1891). The study showed that expression of results relative to a standard reduced between-laboratory variation, and one candidate material was proposed as an International Standard. However, not all study participants agreed with this choice because of concerns about possible instability of the candidate standard when saline was used as diluent in the assay. The Committee recommended that the study authors investigate this concern further. Furthermore, the Committee suggested that the authors consider whether there was a need for product-specific reference materials.

Physicochemical analysis may help to resolve some of the outstanding questions about equivalence of the two formulations currently on the market. The Committee deferred a decision to establish a reference material for botulinum type A toxin and invited the authors to submit revised proposals to a future meeting.

WHO Technical Report Series, No. 897, 2000

Annex 1

Recommendations for the production and control of *Haemophilus influenzae* type b conjugate vaccines[1]

Recommendations published by WHO are intended to be scientific and advisory. Each of the following sections consitutes guidance for national control authorities and for the manufacturers of biological products. If a national control authority so desires, these Recommendations may be adopted as definitive national requirements, or modifications may be justified and made by a national control authority. It is recommended that modifications to these Recommendations be made only on condition that modifications ensure that the vaccine is at least as safe and efficacious as that prepared in accordance with the Recommendations set out below. The parts of each section printed in small type are comments for additional guidance intended for manufacturers and national control authorities which may benefit from those details.

[1] As decided by the WHO Expert Committee on Biological Standardization in October 1998, requirements for biological substances have been renamed "recommendations". These Recommendations are a revised version of the 1991 Requirements for *Haemophilus influenzae* Type b Conjugate Vaccines (Requirements for Biological Substances No. 46).

Introduction

The 1991 Requirements for *Haemophilus influenzae* type b vaccines adopted by the WHO Expert Committee on Biological Standardization in 1990 (*1*) embodied recommendations made by an informal consultation held in 1989. Much experience has been gained with the preparation and control of these vaccines since that time, including the recognition that some vaccine lots that complied with the Requirements had reduced immunogenicity in children (*2*). This emphasized a need to reassess vaccine control strategies and for continued post-marketing surveillance. A revision of the 1991 Requirements (renamed Recommendations in accordance with the decision of the Committee at its forty-ninth meeting, in October 1998) has therefore been prepared, which takes into account the above-mentioned considerations. In particular, it has been shown that the biological assay of potency recommended in 1991 does not correlate with the efficacy of the vaccine in infants and does not provide a sensitive indicator of vaccine quality. Thus, while immunogenicity testing in animals is necessary during vaccine development, the revised Recommendations state that an animal immunogenicity test need not be used for routine batch (lot) release. Instead, the testing focuses on physicochemical tests to monitor consistency of production of the polysaccharide, the protein carrier and the bulk conjugate. Where an *H. influenzae* type b conjugate is included in a combination vaccine, further consideration needs to be given to batch release testing of the final bulk and final fill because of the difficulty in performing physicochemical assays on such preparations.

General considerations

H. influenzae type b causes several diseases in humans, the most common and most serious being meningitis and pneumonia, mainly in children under 5 years of age. Other bacteraemic infections caused by this pathogen include epiglottitis, cellulitis, septic arthritis, osteomyelitis and pericarditis. There are six serotypes of encapsulated *H. influenzae*. Nearly all cases of meningitis and most cases of other bacteraemic diseases are caused by type b organisms where the capsular polysaccharide plays an important role in virulence (*3*). *H. influenzae* type b conjugate vaccines have been developed and shown to be a safe and effective means of protecting against such infections. These Recommendations deal only with *H. influenzae* type b conjugate vaccines, derived from the type b polysaccharide, a linear polymer composed of units of 3-β-D-ribofuranosyl (1→1)-D-ribitol-5-phosphate (referred to as PRP) covalently coupled to a protein carrier (*4*).

A number of epidemiological studies show the public health importance of *H. influenzae* type b disease. In the absence of the general use of the *H. influenzae* type b conjugate vaccine, the annual incidence of disease caused by this bacterium among children in the first 5 years of life is in the range 0.02–0.5% (*5–9*). The disease burden is highest in infants aged between 4 and 18 months, *H. influenzae* disease rarely occurring in infants under 3 months of age or children older than 6 years (*6*). In both developed and developing countries *H. influenzae* type b disease was the leading cause of non-epidemic bacterial meningitis in this age group and, despite prompt and adequate antibiotic treatment, frequently resulted in severe neurological sequelae, the most serious of which are hearing loss and mental retardation (*10, 11*). *H. influenzae* type b disease is contagious and can cause outbreaks of meningitis where susceptible children are crowded together, e.g. in day-care centres (*12, 13*). Transmission of *H. influenzae* occurs by means of droplets expelled by colonized individuals, and asymptomatic carriers are important disseminators of the organism. An additional concern has been the development of strains resistant to antibiotics (*14*), especially ampicillin. Vaccines are the only public health measure available to prevent the majority of disease caused by *H. influenzae* type b.

The low incidence of *H. influenzae* type b disease in older children and adults correlates with the presence of bactericidal antibodies in their serum, directed mainly to the *H. influenzae* type b capsular polysaccharide (*5*). The first vaccine against *H. influenzae* type b disease was therefore made from the type b capsular polysaccharide alone. In 1977, *H. influenzae* type b polysaccharide vaccine was shown to be protective in older children in Finland (*15*). The vaccine was, however, ineffective in inducing antibodies or providing protection in children under 18 months old (*16–19*). This lack of efficacy in the age group at greatest risk stimulated the development of improved *H. influenzae* type b vaccines.

The two immunological properties of the *H. influenzae* type b polysaccharide that limit its use in infants and young children are its age-related immunogenicity, and its failure to elicit immunological memory, and consequently a booster response on subsequent exposure to the polysaccharide (*16–18, 20*). However, immunogenicity can be enhanced by linking the polysaccharide covalently to a protein so that a T-cell-dependent antisaccharide response is elicited (*21–26*). Several different PRP conjugates stimulate T-cell-dependent antibody synthesis in infants and young children with a booster response and, in comparison with polysaccharide vaccine, an increased proportion of IgG antibody relative to IgM antibody (*24, 27–31*). Freeze-dried (lyophilized) and liquid preparations have been licensed for marketing.

Following the introduction of *H. influenzae* type b conjugate vaccines into routine childhood immunization programmes in the 1990s, disease caused by *H. influenzae* type b has largely disappeared in Australia, Canada, western Europe, New Zealand and the United States of America (*32–35*). Efficacy estimates in the range 93–100% have been reported for these vaccines (*28, 36–39*). Indeed, not only do they induce protective circulating antibodies and immunological memory in infants, but their use also results in decreased nasopharyngeal colonization with *H. influenzae* type b. Thus, a herd effect is achieved through reduced transmission of the pathogen (*40–42*).

Special considerations for the production and control of *Haemophilus influenzae* type b conjugate vaccines

The production and control of *H. influenzae* type b conjugate vaccines are more complex than those of unconjugated capsular polysaccharide vaccines, such as meningococcal polysaccharide vaccines (*43*) or pneumococcal polysaccharide vaccines. Polysaccharide vaccines consist of defined chemical entities and, when prepared to the same specifications, are expected to have comparable potencies, regardless of manufacturer. *H. influenzae* type b conjugate vaccines are less well defined chemically and different conjugation chemistries have been shown to yield effective vaccines (*27, 42*). *H. influenzae* type b conjugate vaccines all contain immunogenic determinants capable of stimulating the production of serum IgG antibodies to PRP, and are capable of inducing immunological memory in young children (*24, 26–31, 36*).

Several *H. influenzae* type b conjugate vaccines have been developed and licensed and extensive experience has been gained with the use of some of these products in Europe and the USA. The composition of some of these vaccines is described in Table Al. Immunogenicity in infants, including analysis of IgG subclasses of anti-PRP antibody, can be used to demonstrate equivalence between a new *H. influenzae* type b conjugate vaccine and existing effective vaccines. The existence of a surrogate for protection is important, because it will not be possible to carry out acceptable placebo-controlled protection-based efficacy trials of new *H. influenzae* type b conjugate vaccines or new formulations. Production of specific anti-PRP IgG has been correlated with vaccine efficacy.

Currently available effective vaccines have been found to:

– induce ⩾1 μg/ml of antibody to PRP in 70% or more of infants 1 month after completion of the primary immunization series;

- prime infants for a booster response to the native *H. influenzae* type b polysaccharide given 6–9 months after the primary immunization; and
- produce functional antibodies measured by either opsonic or bactericidal activity.

The immunogenicity in humans of the PRP and protein components of any *H. influenzae* type b conjugate vaccine should be assessed before the vaccine is licensed. Guidance on methods of evaluating immunogenicity is given in Appendix 1.

Currently, no useful biological assay is available to test potency and/or protective efficacy of individual production lots. Consequently, the strategy for control of the vaccine emphasizes the use of tests for molecular characterization and purity to ensure lot-to-lot consistency of composition with the specifications of the vaccine lot(s) employed in definitive clinical trials to prove efficacy and safety. The immunogenicity and induction of memory for antibody production by *H. influenzae* type b conjugate vaccines have been evaluated in mice and guinea-pigs (*23, 44, 45*). However, the results of such evaluations do not correlate with the immunological properties of the vaccines in human infants. Thus, whilst immunogenicity testing in animals is necessary during vaccine development to demonstrate an ability to induce a T-cell-dependent immune response, an animal potency test (immunogenicity) need not be used for routine batch (lot) release. Instead, the testing should focus on physicochemical criteria for monitoring the consistent quality of the polysaccharide, the protein carrier and the bulk conjugate.

Combination vaccines containing a *Haemophilus influenzae* type b conjugate

It is recognized that vaccine combinations may be needed for effective vaccine utilization. Each combination, produced either by preformulation or at the time of reconstitution, should be evaluated by the vaccine producer in the laboratory for possible incompatibilities between the components after mixing and for stability over time, as well as clinically. Because of the difficulty in performing physicochemical assays on *H. influenzae* type b conjugate vaccine in a combination vaccine preparation, further consideration needs to be given to batch release of final bulk and final lot of such vaccines, and the tests applied should be agreed with the national control authority. For example, an immunoassay may be considered in order to confirm that the formulated product consistently retains its immunogenic activity.

H. influenzae type b conjugate vaccines have been successfully combined by formulation with diphtheria–tetanus–pertussis vaccines containing whole-cell pertussis vaccine (DTwP), with hepatitis B vaccine, and with DTwP and inactivated poliomyelitis vaccine (IPV). Such combinations have been demonstrated to be safe and effective. Another type of combination vaccine is formed at the time of use by reconstitution of freeze-dried *H. influenzae* type b conjugate vaccine with, for example, a DTwP vaccine. Combinations with DTwP vaccine have generally shown no incompatibility. Nevertheless, the particular DTwP vaccine to be used should be evaluated clinically for possible adverse interactions with the *H. influenzae* component. A number of clinical studies have demonstrated that combinations of DTP vaccines containing acellular pertussis vaccine (DTaP) and *H. influenzae* type b conjugate vaccine may result in marked diminution of the immune response to the *H. influenzae* polysaccharide (*46*). However, despite reduced immunogenicity, such combination vaccines can prime for immunological memory (*46*).

Table A1
Formulation of some currently available *H. influenzae* type b conjugate vaccines[a,b]

H. influenzae polysaccharide material	Polysaccharide per single human dose (μg)	Nature of carrier protein	Protein per single human dose (μg)
Polysaccharide (size-reduced)	25	Diphtheria toxoid	18
Polysaccharide (low relative molecular mass)	10	Diphtheria CRM 197 protein	25
Polysaccharide (size-reduced)	7.5	Outer membrane protein complex of *Neisseria meningitidis* group B	125
Polysaccharide	10	Tetanus toxoid	20

[a] For guidance only.
[b] *H. influenzae* type b conjugate vaccine is a preparation of capsular polysaccharide from *H. influenzae* type b covalently linked to carrier protein.

Part A. Manufacturing recommendations

A.1 Definitions

A.1.1 *Proper name*

The proper name of the vaccine should be "*Haemophilus influenzae* type b conjugate vaccine" translated into the language of the country of use.

The use of this name should be limited to vaccines that satisfy the recommendations formulated below.

A.1.2 *Descriptive definition*

Haemophilus influenzae type b conjugate vaccine is a preparation of capsular polysaccharide from *H. influenzae* type b covalently linked to a carrier protein.

A.1.3 *International reference materials*

No formally established international reference materials that would allow the standardization of immune responses to *H. influenzae* type b conjugated vaccines are currently available, but their development is under consideration (see also Appendix 1).

A.1.4 *Terminology*

Master seed lot. A suspension of *H. influenzae* type b organisms derived from a strain that has been processed as a single lot and is of uniform composition. It is used for the preparation of working seed lots. Master seed lots should be maintained in the freeze-dried form or frozen at or below −45 °C.

Working seed lot. A suspension of *H. influenzae* type b organisms derived from the master seed lot by growing the organisms and maintaining them in aliquots in the freeze-dried form or frozen state at or below −45 °C. The working seed lot is used, if applicable, after a fixed number of passages, for inoculating production medium.

Single harvest. The material obtained from one batch of cultures that have been inoculated with the working seed lot (or with the inoculum derived from it), harvested and processed together.

Purified polysaccharide. The material obtained after final purification. The lot of purified polysaccharide may be derived from a single harvest or a pool of single harvests processed together.

Processed polysaccharide. Purified polysaccharide that has been modified by chemical reaction or physical process in preparation for conjugation to the carrier.

Carrier. The protein to which processed polysaccharide is covalently linked for the purpose of eliciting a T-cell-dependent immune response to the *H. influenzae* type b polysaccharide.

Bulk conjugate. A conjugate prepared from a single lot or pool of lots of polysaccharide and a single lot of protein or a pool of lots of protein. It is the parent material from which the final bulk is prepared.

Final bulk. The homogeneous preparation present in a single container from which the final containers are filled either directly or through one or more intermediate containers.

Final lot. A number of sealed, final containers that are equivalent with respect to the risk of contamination during filling and, if appropriate, freeze-drying. A final lot should therefore have been filled from a single container and freeze-dried in one continuous working session.

A.2 General manufacturing recommendations

The general manufacturing recommendations contained in Good Manufacturing Practices for Pharmaceutical (*47*) and Biological (*48*) Products should be applied to establishments manufacturing *H. influenzae* type b conjugate vaccines with the addition of the following.

Details of standard operating procedures for the preparation and testing of *H. influenzae* type b conjugate vaccines adopted by a manufacturer, together with evidence of appropriate validation of each production step, should be submitted for approval to the national control authority. All assay procedures used for quality control of the conjugate vaccine and vaccine intermediates should also be validated (*49*). As may be required, proposals for modification of the manufacturing/control methods should also be submitted for approval to the national control authority before they are implemented.

H. influenzae type b is a class 2 pathogen and should be handled under conditions appropriate for this class of microorganism (*50*). Standard operating procedures should be developed for dealing with emergencies involved with accidental spillage, leakage or other dissemination of *H. influenzae* organisms.

Persons employed in the production and control facilities should be adequately trained. Appropriate protective measures, including vaccination against *H. influenzae* type b, should be implemented. Adherence to current good manufacturing practices is important to the integrity of the product, to protect the workers and to protect the environment.

A.3 Production control

A.3.1 *Control of the polysaccharide*

A.3.1.1 Strain of H. influenzae type b

The strain of *H. influenzae* type b to be used in preparing *H. influenzae* type b conjugate vaccine should be identified by a record of its history, including the source from which it was obtained and the tests made to determine the characteristics of the strain. The strain should have been shown to be capable of producing type b polysaccharide.

Nuclear magnetic resonance (NMR) spectroscopy may be used to confirm the identity of the type b polysaccharide produced by the selected *H. influenzae* type b strain.

A.3.1.2 Seed lot system

The production of *H. influenzae* type b polysaccharide should be based on a working seed lot system. If materials of animal origin are used in the medium for seed production, preservation of strain viability for freeze-drying or for frozen storage, then they should comply with the guidance given in the *Report of a WHO Consultation on Medicinal and Other Products in Relation to Human and Animal Transmissible Spongiform Encephalopathies* (*51*) and should be approved by the national control authority. Cultures derived from the working seed lot should have the same characteristics as cultures of the strain from which the master seed lot was derived. Seed lots should be in conformity with the recommendations in section A.3.2.1.

Manufacturers are encouraged to avoid, wherever possible, the use of materials of animal origin.

A.3.1.3 Culture media for production of the polysaccharide

Some media used for the growth of bacteria have been shown to contain human blood-group antigen-like material. Therefore, the *H. influenzae* type b working seed should be inoculated into a liquid medium that does not contain blood-group antigens or high-molecular-weight polysaccharides of either plant or animal origin. Assurance of the absence of human blood-group antigens should be provided by a sensitive immunoassay. This may be done on a lot-to-lot basis or by validation of the process. If materials of animal origin are used, their use should be in accordance with the guidance given in the *Report of a WHO Consultation on Medicinal and Other Products in Relation to Human and Animal Transmissible Spongiform Encephalopathies* (*51*) and should be approved by the national control authority.

Manufacturers are encouraged to avoid, wherever possible, the use of materials of animal origin.

A.3.1.4 Single harvests

Consistency of growth of *H. influenzae* type b microorganisms should be demonstrated by monitoring the growth rate, pH and yield of polysaccharide.

A.3.1.5 Control of bacterial purity

Samples of the culture taken before killing should be tested for contamination. The purity of the culture should be verified by suitable methods that should include inoculation on to appropriate plating media. If any contaminant is found, the culture or any product derived from it should be discarded. The killing process should also be adequately validated.

A.3.1.6 Purified polysaccharide

All steps in the purification procedure should be carried out in clean containers made of material compatible with the solvents in use during the particular step of the procedure. Chemicals of an appropriate purity should be employed.

Each lot of purified polysaccharide should be tested for purity. The limits given below are expressed with reference to the purified polysaccharide, corrected for moisture. Each manufacturer should define limits for its own product, which should be agreed by the national control authority.

> Generally, purified polysaccharide is stored at or below −20 °C to ensure stability.

Identity test. A test should be performed on the purified polysaccharide to verify its identity.

> An immunological method or ^{1}H or ^{13}C NMR spectroscopy (*4*) may be used for this purpose.

Molecular size distribution. The molecular size distribution of each lot of purified polysaccharide should be estimated.

> Separation methods such as gel filtration using cross-linked agarose (e.g. gel filtration using Sepharose CL-4B or CL-2B gels) with a suitable buffer and a refractive index detector (*52*) or high-performance size-exclusion chromatography (HPSEC), either alone or in combination with light scattering and refractive index detectors (e.g. multiple angle laser light scattering; MALLS) are suitable for this purpose (*53*).

The distribution constant (K_D) should be determined by measuring the molecular size distribution of the polysaccharide at the main peak of the elution curve. The K_D value and/or the mass distribution limits should be established and shown to be consistent from lot to lot for a given product.

The K_D value should be determined by gel filtration, the mass distribution limits by MALLS; it is not necessary to do both.

Moisture content. The moisture content of the isolated purified polysaccharide should be determined by a suitable method approved by the national control authority and shown to be consistent within agreed limits.

Polysaccharide composition. The quality of the polysaccharide can be estimated by measuring the ribose content. The ribose content should be not less than 32% of the polysaccharide dry weight (i.e. corrected for moisture), as estimated by the Bial reaction for pentose, using D-ribose as a standard (*54*).

Other methods can be used to define the quantitative composition of the purified polysaccharide (e.g. high-performance anion-exchange chromatography with pulsed amperometric detection (HPAEC-PAD) analysis), but these methods and specifications should be validated for this purpose. If other methods are used, different specifications for composition may apply and should be defined.

The phosphorus content, as determined by the method of Chen et al. (*55*) or another suitable validated method, should be measured and be within defined limits.

The theoretical phosphorus content is 8.4%, and a specification of between 6.8% and 9% calculated on the dry weight has been found suitable. If other methods are used to confirm phosphorus content, direct determination of phosphorus is not required.

Protein impurity. Each lot of purified polysaccharide should contain not more than 1% of protein, calculated on the dry weight and determined by the method of Lowry et al. (*56*), using bovine serum albumin as a reference (*52*), or another suitable validated method.

Sufficient polysacccharide should be assayed to detect accurately 1% contamination.

Nucleic acid impurity. Each lot of purified polysaccharide should contain not more than 1% of nucleic acid calculated on the dry weight and determined by ultraviolet spectroscopy, on the assumption that the absorbance of a nucleic acid solution at a concentration of 10 g/l contained in a cell of 1-cm path length at 260 nm is 200 (*52*), or by another validated method.

Sufficient polysaccharide should be assayed to detect accurately 1% contamination.

Endotoxin content. The endotoxin content of the purified polysaccharide should be determined and shown to be within limits agreed by the national control authority in order to ensure that any pyrogenic activity of the final product is acceptable.

Less than 10 IU of endotoxin per µg of polysaccharide when measured by a *Limulus* amoebocyte lysate test can be achieved. Alternatively, polysaccharide preparations should pass the rabbit pyrogenicity test when injected into rabbits in amounts of 1.0 µg of purified polysaccharide per kg.

A.3.1.7 Processed polysaccharide

Most of the processed polysaccharide preparations are partially depolymerized either before or during the chemical modification.

Chemical modification. Several methods for chemical modification (activation) have been found to be satisfactory. The methods should be approved by the national control authority.

Suitable methods include:

1. The polysaccharide is reacted with cyanogen bromide to introduce groups reactive with "spacer molecules" or with the carrier protein. Excess reactants are removed from the polysaccharide by ultrafiltration or another suitable method.
2. Following cyanogen bromide activation of the polysaccharide, adipic acid dihydrazide is covalently bound to the polysaccharide.
3. Size-reduced polysaccharides are produced by periodate oxidation of the purified polysaccharide. This generates aldehyde groups. The resulting low-molecular-weight PRP is purified.
4. Size-reduced polysaccharides are also produced by controlled acid hydrolysis, which generates reducing-end ribose. After purification, the resulting low-molecular-weight PRP is reductively aminated and converted to the adipic acid spacer *N*-hydroxy-succinimide-activated derivative.
5. The polysaccharide is reacted with carbonyldiimidazole followed by butanediamine to form a reactive intermediate with a terminal amino group. This group is further reacted to form the final derivatized polysaccharide.

The processed polysaccharide to be used in the conjugation reaction should be assessed for the number of functional groups introduced per unit of polysaccharide.

The degree of activation of the chemically modified polysaccharide should be quantitatively assessed to determine the average number of reactive sites per unit of PRP.

Molecular size distribution. The molecular size of the processed polysaccharide will depend on the manufacturing process. The size should be specified for each type of conjugate vaccine and measured by a suitable method, as this may affect the reproducibility of the conjugation process.

The molecular size distribution may be measured by methods such as gel filtration chromatography or HPSEC using MALLS and refractive index (RI) detection (*57, 58*).

A.3.2 *Control of the carrier protein*

A.3.2.1 Microorganisms and culture media for production of the carrier protein

Microorganisms to be used for the production of the carrier protein should be grown in media free from substances likely to cause toxic or allergic reactions in humans. If any materials of animal origin are used in seed preparation or preservation, or in production, their use should be in accordance with the guidance given in the *Report of a WHO Consultation on Medicinal and Other Products in Relation to Human and Animal Transmissible Spongiform Encephalopathies* (*51*) and should be approved by the national control authority.

Production should be based on a seed lot system with the strains identified by a record of their history and of all tests made periodically to verify strain characteristics. Consistency of growth of the microorganisms used should be demonstrated by monitoring the growth rate, pH and yield of appropriate protein(s).

A.3.2.2 Characterization and purity of the carrier protein

Proteins that have been used to date as carriers in *H. influenzae* type b conjugate vaccines include diphtheria and tetanus toxoids, a non-toxic mutant of diphtheria toxin (CRM197) and the outer membrane protein complex of *Neisseria meningitidis* group B (see Table Al). Test methods used to characterize such proteins, to ensure that they are non-toxic, and to determine their purity and concentration should be approved by the national control authority. Physicochemical methods that may be used to characterize such proteins include sodium dodecylsulfate-polyacrylamide gel electrophoresis (SDS-PAGE), isoelectric focusing, high-performance liquid chromatography (HPLC), amino acid analysis, amino acid sequencing, circular dichroism, fluorescence spectroscopy, peptide mapping and mass spectrometry as appropriate (*59*).

Proteins and purification methods that have been used include:

1. Diphtheria and tetanus toxoids. These should be in accordance with the relevant recommendations published by WHO (*60*) and be of high purity. The purity should be, at the very least, 1500 Lf/mg (Lf = limit of flocculation) protein (nondialysable) nitrogen. Purification may

precede detoxification since this may result in a purer product, but particular care should be taken to avoid reversion to toxin when this procedure is used. Amino acids such as lysine are frequently added during detoxification and help prevent reversion. Toxoids may be rendered more suitable for conjugation by additional purification.

2. Diphtheria CRM197 protein, a non-toxic mutant of diphtheria toxin, isolated from cultures of *Corynebacterium diphtheriae* C7/β197 (*22*). Protein of purity greater than 90% as determined by HPLC is prepared by column chromatographic methods. Methods should be in place to distinguish the CRM197 protein from the active toxin.
3. Outer-membrane protein complex of *Neisseria meningitidis* group B extracted from washed bacterial cells with buffer containing detergent. The cell debris is removed and the membrane complex concentrated and washed with the buffer containing detergent to remove impurities. The composition of the purified outer-membrane complex should be determined by SDS-PAGE or a similar method and shown to be consistent from lot to lot, have not more than 8% lipopolysaccharide by weight, and pass the rabbit pyrogenicity test when injected into rabbits in amounts of 0.25 μg/kg of body mass.

A.3.2.3 Processing of carrier protein

In some conjugation procedures, the carrier protein is chemically derivatized before coupling to the polysaccharide. As a measure of consistency, the extent of derivatization of the protein needs to be monitored at this stage.

Protein activation methods that have been used include:

1. The introduction into diphtheria toxoid of a specified concentration of "spacer groups" reactive with activated polysaccharide.
2. The addition of thiol groups to the outer-membrane protein complex of *Neisseria meningitidis* group B.

A.3.3 ***Control of bulk conjugate***

A number of conjugation methods are currently in use (see Table A2); all involve multi-step processes. Both the method and the control procedures used to ensure the reproducibility, stability and safety of the conjugate should be established once the immunogenicity of a particular *H. influenzae* type b conjugate vaccine has been demonstrated. The derivatization and conjugation process should be monitored by analysis for unique reaction products or by other suitable means.

Residual unreacted functional groups potentially capable of reacting in vivo may be present following the conjugation process. The manufacturing process should be validated to show that reactive functional groups do not remain at the conclusion of the manufacturing process.

Table A2

Methods currently used for conjugation of *Haemophilus influenzae* type b polysaccharide and control of conjugates[a]

Method	Procedure	Assay for conjugation
Reductive amination	Combine carrier protein and aldehyde form of polysaccharide in presence of reducing agent	Formation of unique amino acid and gel filtration
Reductive amination and attachment of spacer linked to active ester	Selective reducing end group activation and coupling to carrier protein through spacer	Gel filtration or SDS-PAGE
Carbodiimide-mediated coupling	Combine reactants in presence of carbodiimide	Gel filtration
Cyanogen-bromide activation of polysaccharide	Addition of carrier protein to cyanogen-bromide-activated polysaccharide	Gel filtration and assay for bound polysaccharide
Thioether bonding	Combine haloacyl polysaccharide with protein thiol	Formation of unique amino acid and assay for bound polysaccharide

[a] For guidance only.

After the conjugate has been purified, the tests described below should be performed in order to assess consistency of manufacture. The tests are critical for assuring lot-to-lot consistency.

A.3.3.1 Residual reagents

The conjugate purification procedures should remove residual reagents used in the conjugation process. The removal of reagents, as well as reaction by-products such as cyanide, should be confirmed by suitable tests or by validation of the purification process.

A.3.3.2 Conjugation markers

Where the chemistry of the conjugation reaction results in the creation of a unique linkage marker (e.g. a unique amino acid), each batch should be assessed to quantify the extent of covalent reaction of the PRP with the carrier protein, so that the frequency of the covalent bond is given as a function of the number of PRP repeating units or overall PRP content.

The PRP–protein ratio after purification of the conjugate is another suitable conjugation marker, although not a direct measurement.

A.3.3.3 Residual reactive functional groups

Each batch should be shown to be free of residual reactive functional groups on the chemically modified polysaccharide or the carrier protein, either by analysis of each lot or by validation of the process.

A.3.3.4 Content of PRP

The content of PRP should be chemically determined by means of an appropriate validated assay.

A.3.3.5 Conjugated and unbound (free) PRP

Only the PRP that is covalently bound to the carrier protein, i.e. conjugated PRP, is immunologically important for clinical protection.

Each batch of conjugate should be tested for free, unconjugated PRP in order to ensure that the amount present in the purified bulk is within the limits agreed by the national control authority based on lots shown to be clinically safe and efficacious. Alternatively, the total amount of PRP covalently attached to the carrier protein may be measured after separation from unconjugated PRP.

> A number of methods have been used to separate unconjugated PRP from the conjugate, including precipitation, gel filtration, size-exclusion anion-exchange, and hydrophobic chromatography, ultrafiltration and ultracentrifugation (*61*). The unconjugated PRP can then be quantified by a range of techniques, including HPAEC-PAD and immunoassays with anti-PRP antibodies.
>
> Vaccines demonstrating adequate immunogenicity in clinical studies currently contain amounts of unbound polysaccharide ranging from less than 10% to up to 40% of the total PRP, depending upon the vaccine.

A.3.3.6 Protein content

The protein content of the conjugate should be chemically determined by means of an appropriate validated assay and comply with limits set for that particular product.

> If possible, unconjugated (free) protein should also be measured.

A.3.3.7 PRP-to-protein ratio

The PRP-to-protein ratio of the conjugate should be calculated. For each conjugate, the ratio should be within the range approved for that particular conjugate by the national control authority.

A.3.3.8 Molecular size distribution

The molecular size of the PRP–protein conjugate is an important parameter in establishing consistency of production and in studying physicochemical stability during storage. The relative molecular size of the conjugate should be determined for each bulk using a validated chromatographic method appropriate to the size of the conjugate being evaluated. Suitable methods include gel filtration (for example on Sepharose CL-4B) (*52*) and size-exclusion chromatography (*53*). The method should be shown to distinguish the PRP–protein conjugate from other components that may be present, e.g. unbound protein or polysaccharide.

A.3.3.9 Sterility

The bulk purified conjugate should be tested for bacterial and mycotic sterility in accordance with the recommendations of Part A, sections 5 and 5.2 of the 1973 revised Requirements for the Sterility of Biological Substances (*62*) or by a method approved by the national control authority. If a preservative has been added to the product, appropriate measures should be taken to prevent it from interfering with the test.

A.3.3.10 Specific toxicity of carrier protein in the conjugate

The bulk conjugate should be tested for the absence of specific toxicity of the carrier protein where appropriate (for example, the use of tetanus or diphtheria toxoids).

Absence of specific toxicity of the carrier protein may also be assessed through validation of the production process.

A.3.4 ***Final bulk***

A.3.4.1 Preparation

The final bulk is prepared by mixing the adjuvant and a preservative and/or stabilizer (as appropriate) with a suitable quantity of the bulk conjugate so as to meet the specifications of vaccine lots shown to be safe and efficacious in clinical trials.

A.3.4.2 Sterility

Each final bulk should be tested for bacterial and mycotic sterility as indicated in section A.3.3.9.

A.3.5 ***Filling and containers***

The recommendations concerning filling and containers given in Annex 1, section 4 of Good Manufacturing Practices for Biological Products (*48*) should be applied.

A.3.6 ***Control tests on final product***

A.3.6.1 Identity

An identity test should be performed on at least one labelled container from each final lot.

An immunological test, using antibodies specific for the purified polysaccharide, may be used.

A.3.6.2 Sterility

The contents of final containers should be tested for bacterial and mycotic sterility as indicated in section A.3.3.9.

A.3.6.3 PRP content

The total PRP content in the final containers should be determined; it should be shown to be within the specifications agreed by the national control authority, and within ±20% of the stated PRP content.

The conjugate vaccines produced by different manufacturers differ in formulation (see Table A1). A quantitative assay for the PRP in the final container may be product-specific; colorimetric, chromatographic (including HPLC) or immunological methods may be used (*61*).

A.3.6.4 Residual moisture

If the vaccine is freeze-dried, the average moisture content should be determined by methods accepted by the national control authority. Values should be within the limits for the preparations shown to be adequately stable in the stability studies of the vaccine.

The test should be performed on 1 vial per 1000 up to a maximum of 10 vials but on no less than 5 vials taken at random from throughout the final lot. The average residual moisture content should generally be no greater than 2.5% and no vial should be found to have a residual moisture content of 3% or greater.

A.3.6.5 Pyrogen content

The vaccine in the final container should be tested for pyrogenic activity by intravenous injection into rabbits or by a *Limulus* amoebocyte lysate test. Specifications for endotoxin content or pyrogenic activity should be established; they should be consistent with levels found to be acceptable in vaccine lots used in clinical trials. These should be approved by the national control authority.

Existing *H. influenzae* type b conjugate vaccines pass the rabbit pyrogenicity test when injected into rabbits in amounts of PRP ranging from 0.025 to 1.0 μg per kg of body weight, depending on the protein carrier.

A.3.6.6 Adjuvant content

If an adjuvant has been added to the vaccine, its content should be determined by a method approved by the national control authority.

The amount and nature of the adjuvant should be approved by the national control authority. If aluminium or calcium compounds are used as adjuvants, the amount of aluminium should not exceed 1.25 mg per single human dose and that of calcium 1.3 mg per single human dose.

> The adsorption of the antigen to the adjuvant should be investigated. Consistency of adsorption is important, and the adsorption of production lots should be demonstrated to be within the range of values found for vaccine lots shown to be clinically effective.

A.3.6.7 Preservative content

If a preservative has been added to the vaccine, the content of preservative should be determined by a method approved by the national control authority.

The amount of preservative in the vaccine dose should be shown not to have any deleterious effect on the antigen or to impair the safety of the product in humans. The preservative, its concentration and its limits should be approved by the national control authority.

A.3.6.8 General safety test (innocuity)

Each final lot should be tested for unexpected toxicity (abnormal toxicity) using a test approved by the national control authority.

> This test may be omitted for routine lot release once consistency of production has been well established to the satisfaction of the national control authority and when good manufacturing practices are in place. Each lot, if tested, should pass a test for abnormal toxicity.

A.3.6.9 pH

If the vaccine is a liquid preparation, the pH of each final lot should be tested and shown to be within the range of values found for vaccine lots shown to be safe and effective in clinical trials and stability studies. For a lyophilized preparation, the pH should be measured after reconstitution with the appropriate diluent.

A.3.6.10 Inspection of final containers

Each container in each final lot should be inspected visually, and those showing abnormalities such as improper sealing, lack of integrity and, if applicable, clumping or the presence of particles should be discarded.

A.4 Records

The recommendations in section 8 of Good Manufacturing Practices for Biological Products (*48*) should be applied.

A.5 Retained samples

The recommendations in section 9.5 of Good Manufacturing Practices for Biological Products (*48*) should be applied.

A.6 Labelling

The recommendations in section 7 of Good Manufacturing Practices for Biological Products (*48*) should be applied with the addition of the following.

The label on the carton or the leaflet accompanying the container should indicate:

- the amounts of *H. influenzae* type b polysaccharide (PRP) and protein contained in each single human dose
- the temperature recommended during storage and transport
- if the vaccine is freeze-dried, that after its reconstitution it should be used immediately unless data have been provided to the licensing authority indicating that it may be stored for a limited time
- the volume and nature of the diluent to be added in order to reconstitute a freeze-dried vaccine, specifying that the diluent should be supplied by the manufacturer or approved by the national control authority
- the nature and amount of stabilizer contained in each single human dose (if appropriate)
- the nature and amount of preservative contained in each single human dose (if appropriate).

A.7 Distribution and transport

The recommendations in section 8 of Good Manufacturing Practices for Biological Products (*48*) should be applied.

A.8 Stability testing, storage and expiry date

A.8.1 *Stability testing*

Adequate stability studies form an essential part of vaccine development studies. The stability of the vaccine in its final form and at the recommended storage temperatures should be demonstrated to the satisfaction of the national control authority with final containers from at least three lots of final product from different independent bulk conjugates.

The polysaccharide component of conjugate vaccines is subject to gradual depolymerization at a rate that may vary with the type of conjugate, type of formulation or adjuvant, type of excipients and conditions of storage. The depolymerization can result in reduced molecular size of the PRP component, a reduction in the amount of the PRP bound to the carrier protein (i.e. an increase in free PRP) and a reduced molecular size of the conjugate (*63*). In general, PRP–protein conjugate vaccines are susceptible to gradual depolymerization and the expiry dating needs to be established accordingly.

Tests should be conducted before licensing to determine the extent to which the stability of the product has been maintained throughout the proposed validity period. The free (unconjugated) or conjugated PRP content should be determined as a percentage of the total PRP. The vaccine should meet the recommendations for final product (see Part A, sections 3.6.1 and 3.6.10) up to the expiry date.

Molecular sizing of the final product may be carried out to ensure the integrity of the conjugate.

The desorption of antigen from aluminium-based adjuvants, if used, may take place over time and should be investigated and shown to be within the limits agreed by the national control authority.

Accelerated stability studies may provide additional supporting evidence of the stability of the product but cannot replace real-time studies.

When any changes are made in the production procedure that may affect the stability of the product, the vaccine produced by the new method should be shown to be stable.

The statements concerning storage temperature and expiry date appearing on the label should be based on experimental evidence, which should be submitted for approval to the national control authority.

A.8.2 *Storage conditions*

Storage conditions should be based on stability studies and approved by the national control authority.

Storage of liquid vaccine at a temperature of 2–8 °C has been found to be satisfactory.

Freeze-dried vaccine should be stored at 2–8 °C.

A.8.3 *Expiry date*

The expiry date should be approved by the national control authority and take into consideration the data on the stability of the bulk purified conjugate as well as the results of the stability tests referred to in section A.8.1.

Part B. Recommendations for national control authorities

B.1 General

The general recommendations for control laboratories contained in the Guidelines for National Authorities on Quality Assurance for Biological Products (*64*) should be applied.

B.2 Official release and certification

A vaccine lot should be released only if it fulfils national requirements and/or Part A of these Recommendations.

A statement signed by the appropriate official of the national control authority should be provided at the request of the manufacturing establishments and should certify that the lot of vaccine in question satisfies all national requirements as well as Part A of these Recommendations. The certificate should state the number under which the lot was released by the national controller, and the number appearing on the labels of the containers. Importers of *H. influenzae* type b conjugate vaccines should be given a copy of the official national release document.

The purpose of the certificates is to facilitate the exchange of vaccines between countries. An example of a suitable certificate is given in Appendix 2.

Authors

The previous version of these Recommendations (*1*) was revised following an informal consultation held in Brussels (10–11 April 1997) attended by the following participants:

Dr Y. Arakawa, Director, Department of Bacterial and Blood Products, National Institute of Infectious Diseases, Tokyo, Japan; Dr F. Arminjon, Pasteur Mérieux Connaught, Marcy l'Etoile, France; Dr B. Bolgiano, Division of Bacteriology, National Institute for Biological Standards and Control, Potters Bar, England; Dr N. Borstein, Agence du Médicament, Lyon, France; Dr C. Ceccarini, Chiron Vaccines, Siena, Italy; Dr M. Corbel, Head, Division of Bacteriology, National Institute for Biological Standards and Control, Potters Bar, England; Dr R. Dobbelaer, Head, Biological Standardization, Louis Pasteur Scientific Institute of Public Health, Brussels, Belgium (*Rapporteur*); Dr M. Duchêne, Director, Quality Control and Regulatory Affairs, SmithKline Beecham Biologicals, Rixensart, Belgium; Dr W. Egan, Deputy Director, Office of Vaccine Research and Review, Center for Biologics Evaluation and

Research, Food and Drug Administration, Bethesda, MD, USA; Dr J. Esslinger, Department of Quality Assurance, Swiss Serum and Vaccine Institute, Berne, Switzerland; Dr C. E. Frasch, Chief, Laboratory of Bacterial Polysaccharides, Center for Biologics Evaluation and Research, Food and Drug Administration, Bethesda, MD, USA; Dr E. Fürer, Head, Department of Bacteriology, Swiss Serum and Vaccine Institute, Berne, Switzerland; Dr R.K. Gupta, Wyeth-Lederle Vaccines, New York, NY, USA; Dr J. Hennessey, Director, Bioprocess and Bioanalytical Research, Merck & Co. Inc., West Point, PA, USA; Dr L. Hsieh, Assistant Vice-President, Vaccine Development, Wyeth-Lederle Vaccines, Sanford, NC, USA; Dr A. Kaufhold, SmithKline Beecham Biologicals, Rixensart, Belgium; Dr H. Käyhty, National Public Health Institute, Helsinki, Finland; Dr J.G. Kreeftenberg, Bureau for International Cooperation, National Institute of Public Health and Environmental Protection, Bilthoven, Netherlands; Dr D. Lamb, Pasteur Mérieux Connaught, Swiftwater, PA, USA; Dr P. McVerry, Pasteur Mérieux Connaught, Swiftwater, PA, USA; Dr D. Madore, Wyeth-Lederle Vaccines, West Henrietta, NY, USA; Dr V. Oeppling, Paul Ehrlich Institute, Langen, Germany; Dr G. Orefici, Laboratory of Bacteriology and Medical Mycology, Istituto Superiore di Sanità, Rome, Italy; Dr T. Rogaski-Salter, Merck & Co. Inc., West Point, PA, USA; Dr A. Sabouraud, Pasteur Mérieux Connaught, Marcy l'Etoile, France; Dr M. Schwanig, Paul Ehrlich Institute, Langen, Germany; Dr J. Stalder, Division of Biologicals, Federal Office of Public Health, Berne, Switzerland.

WHO Secretariat

Dr N. Dellepiane, Vaccine Supply and Quality, Global Programme on Vaccines and Immunization, World Health Organization, Geneva, Switzerland; Dr J.L. Di Fabio, Pan American Health Organization/WHO Regional Office for the Americas, Washington, DC, USA; Dr E. Griffiths, Chief, Biologicals, World Health Organization, Geneva, Switzerland; Dr J. Wenger, Childrens Vaccine Initiative/Expanded Programme on Immunization, World Health Organization, Geneva, Switzerland.

The first draft of the revised Recommendations was prepared by Dr C.E. Frasch, Chief, Laboratory of Bacterial Polysaccharides, Center for Biologics Evaluation and Research, Food and Drug Administration, Bethesda, MD, USA and Dr E. Griffiths, Chief, Biologicals, World Health Organization, Geneva, Switzerland.

The revised draft was further discussed at an informal consultation held at the National Institute for Biological Standards and Control, Potters Bar, England, in July 1998, attended by a large number of experts from regulatory agencies, manufacturers and academic institutions.[1]

---

[1] Holliday MR, Jones C. Meeting report: WHO cosponsored workshop on the use of physicochemical methods for the characterization of *Haemophilus influenzae* type b conjugate vaccines. *Biologicals,* 1999, **27**: 51–53.

Acknowledgements

Acknowledgements are due to the following experts for their comments and advice on the draft Recommendations:

Dr Y. Arakawa, Director, Department of Bacterial and Blood Products, National Institute of Infectious Diseases, Tokyo, Japan; Dr V.V. Bencomo, Head, Laboratory Synthetic Antigens, University of Havana, Havana, Cuba; Dr B. Bolgiano, Division of Bacteriology, National Institute for Biological Standards and Control, Potters Bar, England; Dr D. Calam, European Coordinator, National Institute for Biological Standards and Control, Potters Bar, England; Mr P. Castle, Principal Scientific Officer, European Pharmacopoeia, Strasbourg, France; Dr M. Corbel, Head, Division of Bacteriology, National Institute for Biological Standards and Control, Potters Bar, England; Dr M. de los Angeles Cortés Castillo, Director, Quality Control, National Institute of Hygiene, Mexico City, Mexico; Dr M. Duchêne, Director, Technical Affairs, SmithKline Beecham Biologicals, Rixensart, Belgium; Dr W. Egan, Deputy Director, Office of Vaccine Research and Review, Center for Biologics Evaluation and Research, Food and Drug Administration, Bethesda, MD, USA; Dr C.E. Frasch, Chief, Laboratory of Bacterial Polysaccharides, Center for Biologics Evaluation and Research, Food and Drug Administration, Bethesda, MD, USA; Dr B.D. Garfinckle, Vice-President, Vaccine Quality Operations, Merck & Co. Inc., West Point, PA, USA; Professor I. Gust, Director, Research and Development, CSL Ltd, Parkville, Australia; Dr J.P. Hennessey, Director, Bioprocess and Bioanalytical Research, Merck & Co. Inc., West Point, PA, USA; Dr C.L. Hsieh, Assistant Vice-President, Vaccine Development, Wyeth-Lederle Vaccines, Sanford, NC, USA; Dr C. von Hunoltstein, Laboratory of Bacteriology and Medical Mycology, Istituto Superiore di Sanità, Rome, Italy; Dr S.S. Jadhav, Executive Director, Serum Institute of India Ltd, Pune, India; Dr C. Jones, Head, Laboratory for Molecular Structure, National Institute for Biological Standards and Control, Potters Bar, England; Dr J.G. Kreeftenberg, Bureau of International Affairs, National Institute of Public Health and Environmental Protection, Bilthoven, Netherlands; Dr V. Oeppling, Paul Ehrlich Institut, Langen, Germany; Dr N. Ravenscroft, Chiron Vaccines, Siena, Italy; Dr J.B. Robbins, National Institutes of Health, Bethesda, MD, USA; Dr J.C. Vincent-Falquet, Director, Scientific Affairs, Pasteur Mérieux Connaught, Marcy l'Etoile, France.

References

1. Requirements for *Haemophilus* Type b Conjugate Vaccines. In: *WHO Expert Committee on Biological Standardization.* Forty-first report. Geneva, World Health Organization, 1991, Annex 1 (WHO Technical Report Series, No. 814).
2. **Egan W, Frasch CE, Anthony BF.** Lot-release criteria, postlicensure quality control and the *Haemophilus influenzae* type b conjugate vaccines. *Journal of the American Medical Association*, 1995, **273**:888–889.
3. **Pittman M.** Variation and type specificity in the bacterial species *Haemophilus influenzae. Journal of Experimental Medicine*, 1931, **53**:471–492 .
4. **Crisel RM et al.** Capsular polymer of *Haemophilus influenzae* type b. *Journal of Biological Chemistry,* 1975, **250**:4926–4933.
5. **Fothergill LD, Wright J.** Influenzal meningitis: relation of age incidence to the bactericidal power of blood against the causal organisms. *Journal of Immunology,* 1933, **24**:273–284 .
6. **Parke JC et al.** The attack rate, age incidence, racial distribution, and case-fatality rate of *Haemophilus influenzae* type b meningitis in Mecklenburg County, North Carolina. *Journal of Pediatrics,*1972, **81**:765–769.
7. **Ward JI et al.** *Haemophilus influenzae* in Alaskan Eskimos: characteristics of a population with an unusual incidence of invasive diseases. *Lancet*, 1981, i:121–125.
8. **Claesson B et al.** Incidence and prognosis of *Haemophilus influenzae* meningitis in children in a Swedish region. *Pediatric Infectious Diseases,* 1984, **3**:36–39.
9. **Cochi SL, Broome CV, Hightower AW.** Immunization of US children with *Haemophilus influenzae* type b polysaccharide vaccine. A cost-effectiveness model of strategy assessment. *Journal of the American Medical Association,* 1985, **253**:521–529.
10. **Sell SH.** Long-term sequelae of bacterial meningitis in children. *Pediatric Infectious Diseases,* 1983, **2**:90–93.
11. **Lindberg J et al.** Long-term outcome of *Haemophilus influenzae* meningitis related to antibiotic treatment. *Pediatrics*, 1977, **60**:1–6.
12. **Glode MP et al.** An outbreak of *Haemophilus influenzae* type b meningitis in an enclosed hospital population. *Journal of Pediatrics*, 1976, **88**:36–40.
13. **Redmond SR, Pichichero ME.** *Haemophilus influenzae* type b disease. An epidemiologic study with special reference to day-care centres. *Journal of the American Medical Association,* 1984, **252**:2581–2584.
14. **Peter G.** Treatment and prevention of *Haemophilus influenzae* type b meningitis. *Pediatric Infectious Diseases,* 1987, **6**:787–790.
15. **Peltola H et al.** *Haemophilus influenzae* type b capsular polysaccharide vaccine in children: a double blind study of 100 000 vaccinees 3 months to 5 years of age in Finland. *Pediatrics*, 1977, **60**:730–737.
16. **Centers for Disease Control.** Polysaccharide vaccine for prevention of *Haemophilus influenzae* type b disease. *Morbidity and Mortality Weekly Report,* 1985, **34**:210–215.
17. **Parke JC et al.** Interim report of a field trial of immunization with capsular polysaccharides of *Haemophilus influenzae* type b and group C *Neisseria*

meningitidis in Mecklenburg county, North Carolina (March 1974–March 1976). *Journal of Infectious Diseases,* 1977, **136**:S51–S56.

18. Robbins JB et al. Quantitative measurement of "natural" and immunization-induced *Haemophilus influenzae* type b capsular polysaccahride antibodies. *Pediatric Research*, 1973, 7:103–110.
19. Smith DH et al. Responses of children immunized with the capsular polysaccharide of *Haemophilus influenzae* type b. *Pediatrics,* 1973, **52**:637–644.
20. Käyhty H et al. Serum antibodies after vaccination with *Haemophilus influenzae* type b capsular polysaccharide: no evidence of immunologic tolerance or memory. *Pediatrics,* 1984, **74**:857–865.
21. Schneerson R et al. Preparation, characterization and immunogenicity of *Haemophilus influenzae* type b polysaccharide–protein conjugates. *Journal of Experimental Medicine,* 1980, **152**:361–376.
22. Anderson P. Antibody responses to *Haemophilus influenzae* type b and diphtheria toxin induced by conjugates of oligosaccharides of the type b capsule with the nontoxic protein CRM197. *Infection and Immunity,* 1983, **39**:233–238.
23. Chu CY et al. Further studies on the immunogenicity of *Haemophilus influenzae* type b and pneumococcal type 6A polysaccharide-protein conjugates. *Infection and Immunity,* 1983, **40**:245–256.
24. Anderson PW et al. Vaccines consisting of periodate-cleaved oligosaccharides from the capsule of *Haemophilus influenzae* type b couples to a protein carrier: structural and temporal requirements for priming in the human infant. *Journal of Immunology,* 1986, **137**:1181–1186.
25. Marburg S et al. Bimolecular chemistry of macromolecules — synthesis of bacterial polysaccharide conjugates with *Neisseria meningitidis* membrane protein. *Journal of the American Chemical Society,* 1986, **108**:5282-5287.
26. Schneerson R et al. Vaccines composed of polysaccharide–protein conjugates: current status, unanswered questions, and prospects for the future. In: Bell R, Torrigiani G, eds. *Towards better carbohydrate vaccines.* Chichester, Wiley, 1987:307–327 (published on behalf of the World Health Organization).
27. Madore DV et al. Safety and immunogenicity of *Haemophilus influenzae* type b oligosaccharide/CRM197 conjugate vaccine in children aged 15 to 23 months. *Pediatrics*, 1990, **86**:527–534.
28. Eskola J et al. Efficacy of *Haemophilus influenzae* type b polysaccharide–diphtheria toxoid conjugate vaccine in infancy. *New England Journal of Medicine,* 1987, **317**:717–722.
29. Einhorn MS et al. Immunogenicity in infants of *Haemophilus influenzae* type b polysaccharide in a conjugate vaccine with *Neisseria meningitidis* outer-membrane protein. *Lancet,* 1986, ii:299–302.
30. Claesson BA et al. Clinical and immunological responses to the capsular polysaccharide of *Haemophilus influenzae* type b alone or conjugated to tetanus toxoid in 18- to 23-month-old children. *Journal of Pediatrics,* 1988, **112**:695–702.
31. Greenberg DP et al. Protective efficacy of *Haemophilus influenzae* type b polysaccharide and conjugate vaccines in children 18 months of age and older. *Journal of the American Medical Association,* 1991, **265**:987–992.

32. **Peltola H, Kilpi T, Anttila M.** Rapid disappearance of *Haemophilus influenzae* type b meningitis after routine childhood immunisation with conjugate vaccines. *Lancet,* 1992, **340**:592–594.

33. **Black SB et al.** Immunization with oligosaccharide conjugate *Haemophilus influenzae* type b (HbOC) vaccine of a large health maintenance organization population: extended follow-up and impact on *Haemophilus influenzae* disease epidemiology. *Pediatric Infectious Disease Journal,* 1992, **11**:610–613.

34. **Slack MPE.** Invasive *Haemophilus influenzae* disease: the impact of Hib immunisation. *Journal of Medical Microbiology,* 1995, **42**:75–77.

35. **Centers for Disease Control.** Progress toward elimination of *Haemophilus influenzae* type b disease among infants and children — United States, 1987–1995. *Mortality and Morbidity Weekly Report,* 1996, **45**:901–906.

36. **Centers for Disease Control.** Update: prevention of *Haemophilus influenzae* type b disease. *Mortality and Morbidity Weekly Report,* 1987, **37**:13–16.

37. **Frasch CE.** Regulatory perspectives in vaccine licensure. In: Ellis RW, Granoff DM, eds. *Development and clinical uses of Haemophilus b conjugate vaccines.* New York, Marcel Dekker, 1994:435–453.

38. **Makela PH et al.** Clinical experience with *Haemophilus influenzae* type b conjugate vaccines. *Pediatrics*, 1990, **85**(suppl.):651–653.

39. **Booy R et al.** Efficacy of *Haemophilus influenzae* type b conjugate vaccine PRP-T. Lancet, 1994, **344**:362–366.

40. **Adegbola RA et al.** Vaccination with *Haemophilus influenzae* type b conjugate vaccine reduces oropharyngeal carriage of *H. influenzae* type b among Gambian children. *Journal of Infectious Diseases*, 1998, **177**:1758–1761.

41. **Takala AK et al.** Reduction of oropharyngeal carriage of *Haemophilus influenzae* type b (Hib) in children immunized with an Hib conjugate vaccine. *Journal of Infectious Diseases*, 1991, **164**:982–986.

42. **Mäkelä PH et al.** Vaccines against *Haemophilus influenzae* type b. In: AlaAladeen DAA, Hormaeche CE, eds. *Molecular and clinical aspects of bacterial vaccine development.* Chichester, Wiley,1995:41–91.

43. Requirements for meningococcal polysaccharide vaccine (Requirements for Biological Substances No. 23). In: *WHO Expert Committee on Biological Standardization. Twenty-seventh report.* Geneva, World Health Organization, 1976, Annex 2 (WHO Technical Report Series, No. 594).

44. **Siber GR et al.** Development of a guinea-pig model to assess immunogenicity of *Haemophilus influenzae* type b capsular polysaccharide conjugate vaccines. *Vaccine*, 1995, **13**:525–531.

45. **Gupta RK et al.** Development of a guinea-pig model for potency/immunogenicity evaluation of diphtheria, tetanus, acellular pertussis (DtaP) and *Haemophilus influenzae* type b polysaccharide conjugate vaccines. *Developments in Biological Standardization,* 1996, **86**:283–296.

46. **Goldblatt D et al.** The induction of immunologic memory after vaccination with *Haemophilus influenzae* type b conjugate and acellular pertussis-containing diphtheria, tetanus and pertussis vaccine combination. *Journal of Infectious Diseases*, 1999, **180**:538–541.

47. Good manufacturing practices for pharmaceutical products. In: *WHO Expert Committee on Specifications for Pharmaceutical Preparations. Thirty-second*

report. Geneva, World Health Organization, 1992, Annex 1 (WHO Technical Report Series, No. 823).

48. Good manufacturing practices for biological products. In: *WHO Expert Committee on Biological Standardization. Forty-second report.* Geneva, World Health Organization, 1992, Annex 1 (WHO Technical Report Series, No. 822).
49. Validation of analytical procedures used in the examination of pharmaceutical materials. In: *WHO Expert Committee on Specifications for Pharmaceutical Preparations. Thirty-second report.* Geneva, World Health Organization, 1993, Annex 5 (WHO Technical Report Series, No. 823).
50. *Biosafety guidelines for personnel engaged in the production of vaccines and biological products for medical use.* Geneva, World Health Organization, 1995 (unpublished document WHO/CDS/BVI/95.5, available on request from Communicable Diseases, World Health Organization, 1211 Geneva 27, Switzerland).
51. *Report of a WHO Consultation on Medicinal and Other Products in Relation to Human and Animal Transmissible Spongiform Encephalopathies, Geneva, 24–26 March 1997.* Geneva, World Health Organization, 1997 (unpublished document WHO/EMC/ZOO/97.3 – WHO/BLG/97.2; available on request from Quality Assurance and Safety: Biologicals, World Health Organization, 1211 Geneva 27, Switzerland).
52. Requirements for meningococcal polysaccharide vaccine (Addendum 1980). In: *WHO Expert Committee on Biological Standardization. Thirty-first report.* Geneva, World Health Organization, 1981, Annex 6 (WHO Technical Report Series, No. 658).
53. Hennessey JP et al. Molecular size analysis of *Haemophilus influenzae* type b capsular polysaccharide. *Journal of Liquid Chromatography*, 1993, **16**:1715–1792.
54. Kabat EA, Mayer MM. *Experimental immunology.* Springfield, IL, Charles C. Thomas, 1961:530.
55. Chen PS et al. Microdetermination of phosphorus. *Analytical Chemistry,* 1956, **28**:1756–1758.
56. Lowry OH et al. Protein measurement with the folin phenol reagent. *Journal of Biological Chemistry*, 1951, **193**:265–275.
57. Ravenscroft N et al. Size determination of bacterial capsular oligosaccharides for their use in conjugate vaccines, *Vaccine*, 1999, **17**:2802–2816.
58. Constantino P et al. Size fractionation of bacterial capsular polysaccharides for their use in conjugate vaccines, *Vaccine*, 1999, **17**:1251–1263.
59. Crane DT, Bolgiano B, Jones C. Comparison of the diphtheria mutant toxin, CRM197, with a *Haemophilus influenzae* type b polysaccharide–CRM197 conjugate by optical spectroscopy. *European Journal of Biochemistry*, 1997, **246**:320–327.
60. Requirements for diphtheria, tetanus, pertussis and combined vaccines. In: *WHO Expert Committee on Biological Standardization. Fortieth report.* Geneva, World Health Organization, 1990, Annex 2 (WHO Technical Report Series, No. 800).
61. Tsai C-M, Gu XX, Byrd BA. Quantification of polysaccharide in *Haemophilus influenzae* type b conjugate and polysaccharide vaccines by high-performance anion-exchange chromatography with pulsed amperometric detection. *Vaccine,* 1994, **12**:700–706.

62. General Requirements for the Sterility of Biological Substances. In: *WHO Expert Committee on Biological Standardization. Twenty-fifth report.* Geneva, World Health Organization, 1973, Annex 4 (WHO Technical Report Series, No. 530).

63. Sturgess A et al. *Haemophilus influenzae* type b conjugate vaccine stability: catalytic depolymerization of PRP in the presence of aluminium hydroxide. *Vaccine*, 1999, **17**:1169–1178.

64. Guidelines for national authorities on quality assurance for biological products. In: *WHO Expert Committee on Biological Standardization. Forty-second report.* Geneva, World Health Organization, 1992, Annex 2 (WHO Technical Report Series, No. 822).

Appendix 1

Evaluation of immunogenicity of *Haemophilus influenzae* type b conjugate vaccines in humans

Different lots of *Haemophilus influenzae* type b conjugate vaccines from each manufacturer should be evaluated for immunogenicity in the target age group before licensing.

Response to polysaccharide

The results of the radioimmunoassay (RIA) of antibodies to the polysaccharide (PRP) (*1,2*) have been shown to be the best available correlate of the clinical effectiveness of *H. influenzae* type b vaccines. One method of assessing the immunogenicity is to assay the serum antibody response to the PRP using a suitable RIA or demonstrated equivalent enzyme-linked immunosorbent assay (ELISA) in which an *H. influenzae* type b reference human serum is included (*3, 4*). This applies to serum samples taken usually just before each injection of a vaccine dose and 1 month after the final dose of the primary series or after a booster dose. The percentage of vaccinees with a serum antibody concentration equal to, or greater than, 0.15 μg/ml and 1.0 μg/ml should be reported, as well as the geometric mean antibody concentration and the distribution of values. It is important to examine the duration of the protective antibody response up to 4 years of age. It is also important to show that the conjugate vaccine stimulates a statistically significant increase in anti-PRP IgG response. However, the concentration of serum antibodies sufficient to confer protection following the use of *H. influenzae* type b conjugate vaccines, or for eliminating carriage, is unclear (*5–8*). Immunological memory is thought to be a significant part of the protective response to the *H. influenzae* type b conjugate vaccine (*9*) and the demonstration of a booster response to native polysaccharide is considered an important part of the evidence of a protective response. The avidity of the induced antibodies is also another area of interest (*6–8*).

The functional activity of the conjugate-induced antibodies should be assessed by measuring the serum bactericidal activity against *H. influenzae* type b (*1, 5*). ELISAs have been used to measure total anti-PRP content, as well as isotype and subclass composition.

Response to carrier protein

Serum antibodies to the carrier protein should be measured in recipients of *H. influenzae* type b conjugate vaccine to ensure that the conjugate vaccine does not interfere with protective immunity that is relevant to that protein. In addition, the measurement provides information about the potential of the *H. influenzae* type b conjugate to serve as a dual immunogen for both the *H. influenzae* type b polysaccharide and the carrier protein. To date, proteins such as diphtheria and tetanus toxoids and an outer membrane protein complex of *Neisseria meningitidis* group B have been used in the preparation of *H. influenzae* type b conjugates. Since some of these carriers are also components of other infant vaccines (e.g. DTP), antibody responses to those vaccines should be measured to ensure that there is no immune interference of clinical relevance. The assay for these antibodies should be a bioassay or an established equivalent.

The following reagents are available through the courtesy of manufacturers and national control authorities:

- *H. influenzae* type b reference serum USA/FDA lot 1983 with 70 μg/ml of total anti-Hib polysaccharide antibody, available from the Center for Biologics Evaluation and Research, Food and Drug Administration, HFM-428, 1401 Rockville Pike, Rockville, MD 10852-1442, USA.
- Short-chained *H. influenzae* type b polysaccharide–serum albumin conjugate (Hb0-HA antigen) prepared by Wyeth-Lederle Vaccines and available from the National Institute for Biological Standards and Control, Potters Bar, Herts. EN6 3QG, England.

References

1. Robbins JB et al. Quantitative measurement of "natural" and immunization-induced *Haemophilus influenzae* type b capsular polysaccharide antibodies. *Pediatric Research,* 1973, **7**:103–110.
2. Kuo JS et al. A radioactive antigen-binding assay for measurement of antibody to *Haemophilus influenzae* type b capsular polysaccharide. *Journal of Immunological Methods,* 1981, **43**:35–47.
3. Phipps DC et al. An ELISA employing a *Haemophilus influenzae* type b oligosaccharide–human serum albumin conjugate correlates with the radioantigen binding assay. *Journal of Immunological Methods,* 1990, **135**:121–128.
4. Madore DV et al. Interlaboratory study evaluating quantitation of antibodies to *Haemophilus influenzae* type b polysaccharide by enzyme-linked immunosorbent assay. *Clinical and Diagnostic Laboratory Immunology*, 1996, **3**:84–88.
5. Smith DH et al. Responses of children immunized with the capsular polysaccharide of *Haemophilus influenzae* type b. *Pediatrics*, 1973, **52**:637–641.
6. Schlesinger Y et al. Avidity and bactericidal activity of antibody elicited by different *Haemophilus influenzae* type b conjugate vaccines. *Journal of the American Medical Association,* 1992, **267**:1489–1494.

7. **Käyhty H.** Difficulties in establishing a serological correlate of protection after immunization with *Haemophilus influenzae* conjugate vaccines. *Biologicals,*1994, **22**:397–402.
8. **Mäkelä PH et al.** Vaccines against *Haemophilus influenzae* type b. In: Ala'Aladeen DAA, Hormaeche CE, eds. *Molecular and clinical aspects of bacterial vaccine development.* Chichester, Wiley, 1995:41–91.
9. **Goldblatt D et al.** The induction of immunologic memory after vaccination with *Haemophilus influenzae* type b conjugate and acellular pertussis-containing diphtheria, tetanus and pertussis vaccine combination. *Journal of Infectious Diseases,* 1999, **180**:538–541.

Appendix 2

Model certificate for the release of *Haemophilus influenzae* type b conjugate vaccines[1]

The following lots of *H. influenzae* type b conjugate vaccine produced by __________________,[2] whose numbers appear on the labels of the final containers, meet all national requirements,[3] and Part A of the Recommendations for the Production and Control of *H. influenzae* type b Conjugate Vaccines,[4] and comply with Good Manufacturing Practices for Pharmaceutical[5] and Biological Products.[6]

Lot no.	Expiry date	Lot no.	Expiry date
________	________	________	________
________	________	________	________
________	________	________	________

As a minimum, this certificate is based on an examination of the manufacturing protocol.

The number of this certificate is ________________________________

The Director of the National Control Laboratory (or Authority as appropriate):[7]

Name (typed) ________________________________

Signature ________________________________

Date ________________________________

[1] To be completed by the national control authority of the country where the vaccine has been manufactured, and to be provided by the vaccine manufacturer to importers.

[2] Name of manufacturer.

[3] If any national requirement(s) is (are) not met, specify which one(s) and indicate why release of the lot(s) has nevertheless been authorized by the national control authority.

[4] With the exception of the provisions on shipping, which the national control authority may not be in a position to control.

[5] WHO Technical Report Series, No. 823, 1992, Annex 1.

[6] WHO Technical Report Series, No. 822, 1992, Annex 1.

[7] Or his or her representative.

WHO Technical Report Series, No. 897, 2000

Annex 2

1990 Requirements for poliomyelitis vaccine (oral) (Addendum 1998)

(Requirements for Biological Substances No. 45)

Introduction

Since the adoption of the 1990 Requirements for Poliomyelitis Vaccine (Oral) (*1*), international collaborative research has identified the molecular mechanisms and genetic determinants of attenuation and reversion of the Sabin poliovirus strains used for manufacture of oral poliomyelitis vaccine (OPV) (*2*). Evidence strongly suggests that mutations in the 5′ non-coding region of the poliovirus genome are critical for the attenuated phenotype, especially for the Sabin type 3 strain. A new molecular technique, called mutant analysis by polymerase chain reaction and restriction enzyme cleavage (MAPREC) (*3*), was developed to quantify reversion of the key mutation(s). For example, a change from "U" to "C" at nucleotide 472 in more than about 1% of genomes in a batch of Sabin poliovirus type 3 is strongly associated with failure in the monkey neurovirulence test. Experience with the MAPREC assay was reviewed at a WHO informal consultation in 1997, which concluded that the assay was a sensitive, robust and standardized molecular biological assay suitable for use by manufacturers and national control laboratories for quality control of OPV (*4*). The consultation also noted that MAPREC is the first example of a new class of tests for the molecular consistency of production of live virus vaccine. An International Standard (WHO Technical Report Series, No. 878) and two International Reference Reagents (WHO Technical Report Series, No. 889) have been established by the WHO Expert Committee for Biological Standardization for MAPREC assay of Sabin poliovirus type 3 and are available from the National Institute for Biological Standards and Control, Potters Bar. This addendum is intended, first of all, to introduce the MAPREC assay as an additional test in the 1990 Requirements for Poliomyelitis Vaccine (Oral).

Second, safety issues related to simian virus 40 (SV40) and poliomyelitis vaccines were considered at a WHO informal consultation in 1997 (*5*). The meeting was convened to examine the effectiveness in excluding SV40 of tests for adventitious agents specified in the 1990 Requirements (*1*). This was in response to renewed interest in SV40 as a possible human pathogen and the availability of new molecular assays to screen for it. Results with the new assays were presented at the meeting and showed that the existing tests for adventitious agents had effectively excluded SV40 from OPV for over 30 years. While this was reassuring, the consultation considered that an additional level of security could be provided by ensuring that SV40 sequences were absent from poliovirus seed stocks. Reference materials for the detection of SV40 sequences are available from the National Institute for Biological Standards and Control, Potters Bar. This addendum also introduces a test to show, by molecular methods, that poliovirus seed stocks are free of SV40 sequences.

Primary monkey kidney cells remain an acceptable and widely used cell substrate for the production of OPV, provided that, as with other cell substrates, the current requirements to exclude viral contaminants are fully implemented. Since the adoption of the 1990 Requirements, there has been a trend towards the use of colony-bred or intensively monitored animals to derive cells for production. This addendum offers guidance to encourage this approach.

Great progress has recently been made towards the eradication of wild-type polioviruses, and the goal of eradicating poliomyelitis is in sight (*6*). Once wild-type polioviruses are eradicated, laboratories will be the only remaining source of the virus. Safe handling and, ultimately, maximum containment of poliovirus in the laboratory is crucial. A systematic, worldwide plan of action to prevent transmission of wild poliovirus from the laboratory to the community was developed by a WHO Working Group (*7*). The Group recommended that the 1990 Requirements for Poliomyelitis Vaccine (Oral) be reviewed to assess the need for wild polioviruses in control assays. While retaining for the time being control assays with wild-type polioviruses, this addendum introduces, as an interim measure, proposals for enhanced laboratory containment protocols.

General considerations

Section 3.2.2 of Part A (*1*) has been revised to exclude SV40 sequences from poliovirus seed lots.

The MAPREC assay for Sabin poliovirus type 3 is being used by a number of manufacturers and national control laboratories to characterize virus seed lots and to monitor the consistency of manufacture of monovalent vaccine bulks. Section 4.4.5.2 of Part A (*1*) has been revised to provide guidance, based on the consensus of a WHO informal consultation (*4*), on the use of the MAPREC test. The addition is reproduced in small print because the database of MAPREC results for existing manufacturers' products is not yet complete, and because further discussions are required on strategies for use of the test that were proposed but not agreed at the 1997 consultation. The use of the MAPREC assay is encouraged, as it has the potential to reduce the use of laboratory animals.

Section 4.4.5.2 of Part A (*1*) has been revised to reflect the need to improve laboratory containment of wild polioviruses.

Section 4.1 of Part C (*1*) has been revised to encourage the use of colony-bred or intensively monitored animals as source of primary kidney-cell cultures, and to offer relevant guidance.

Part A. Manufacturing requirements

3.2 Virus strains

3.2.2 *Tests on virus seed lots*

Replace the first sentence of section 3.2.2 with the following:

"The virus working seed lot used for the production of vaccine batches shall be free from detectable extraneous viruses, free from detectable SV40 sequences, and shall satisfy the requirements specified in Part A, sections 4.3 and 4.4.

Sequences of SV40 are widely used as molecular biological reagents, and contamination of polymerase chain reaction (PCR) assays is potentially a major problem. One approach is to identify separate genomic regions of SV40 for amplification, and to use one region for screening purposes and the other for the confirmation of repeatedly positive samples. It is useful if the second genomic region used for confirmation varies between isolates of different sources, as it is then possible to show that the second genomic region has a unique sequence and that positive results are not due to contamination with laboratory strains of SV40. The sensitivity of the PCR assays for the genomic regions used should be established."

(The remainder of this section is unchanged.)

4.4 Control of bulk suspension

4.4.5.2 Tests in vitro

Replace the first paragraph of section 4.4.5.2 with the following:

"The virus in the filtered bulk suspension shall be tested by at least one in vitro test.

For Sabin poliovirus type 3, the MAPREC assay is suitable as a test of consistency of production; guidance is given below on the use of the assay.

The MAPREC assay should be performed for nucleotide 472-C according to the standard operating procedure developed in international collaborative studies (protocol available from Coordinator, Quality Assurance and Safety: Biologicals, World Health Organization, 1211 Geneva 27, Switzerland), or according to a validated alternative procedure. Results should be expressed as ratios relative to the International Standard for MAPREC Analysis of Poliovirus Type 3 (Sabin). A poliovirus type 3 filtered bulk suspension found to have significantly more 472-C nucleotide than the International Standard (as specified in the standard operating procedure) fails the MAPREC assay. Since the assay is highly predictive of in vivo neurovirulence, a filtered bulk suspension failing the MAPREC assay should not be tested for neurovirulence in monkeys and should not be used to prepare vaccine. Such a failure should trigger an evaluation of the consistency of the manufacturing process, including the suitability of the working seed virus. Filtered bulk suspensions passing the MAPREC assay must be subsequently tested for neurovirulence in monkeys.

The MAPREC assay may also be useful for tests of consistency of single harvests prior to pooling and preparation of the filtered bulk suspension.

The MAPREC test for poliovirus type 3 is also extremely useful for process development, when, for example, an established manufacturer changes production conditions or a new manufacturer starts production. However, in such situations the product also needs to be validated by the monkey neurovirulence test.

Manufacturers are encouraged to gain experience with the MAPREC assay to allow for a smooth transition after the eventual replacement of one or more of the tests on bulk suspensions in the existing requirements with the MAPREC assay.

The virus in the filtered bulk suspension shall be tested for the property of reproducing at temperatures of 36 °C and 40 °C in comparison with the seed lot or a reference virus preparation for the marker tests and with appropriate rct/40– and rct/40+ strains of poliovirus of the same type. Wild polioviruses, defined as field isolates or reference strains derived from polioviruses known or believed to have circulated persistently in the community, that are used as rct/40+ controls in this test must be contained within the laboratory at progressively higher levels of containment in accordance with the global action plan and timetable for safe handling of wild polioviruses (*7*).

The use of suitable vaccine-derived strains, defined as progeny of oral poliovirus vaccine strains, as rct/40+ controls is encouraged.

The incubation temperatures used in this test shall be controlled to within +/− 0.1 °C."

(The rest of this section is unchanged.)

Part C. Requirements for poliomyelitis vaccine (oral) prepared in primary cultures of monkey kidney cells

4.1 Control of source materials

4.1.1 *Monkeys used for preparation of kidney-cell cultures and for testing of virus*

At the end of the first paragraph of section 4.1.1 add the following:

> "Animals from closed or intensively monitored colonies should be used wherever possible."

(The rest of the section is unchanged until the last paragraph.)

Immediately before the final paragraph (in small print) insert the following:

> "It is desirable for kidney-cell cultures to be derived from monkeys shown to be free from antibodies to foamy viruses."

References

1. Requirements for poliomyelitis vaccine (oral). (Requirements for Biological Substances No. 7) (Revised 1989). In: *WHO Expert Committee on Biological Standardization. Fortieth report.* Geneva, World Health Organization, 1990 (WHO Technical Report Series, No. 800).
2. Wood DJ, Macadam AJ. Laboratory tests for live attenuated poliovirus vaccines. *Biologicals*, 1997, **25**:3–15.
3. Chumakov KM et al. Correlation between amount of virus with altered nucleotide sequence and the monkey test for acceptability of oral poliovirus vaccine. *Proceedings of the National Academy of Sciences of the United States of America*, 1991, **88**:199–203.
4. *Report of a WHO informal consultation on the development and use of the MAPREC assay in the quality control of oral poliovirus vaccine, Geneva, September 17–18, 1997.* Geneva, World Health Organization, 1997 (unpublished document; available on request from Quality Assurance and Safety: Biologicals, World Health Organization, 1211 Geneva 27, Switzerland).
5. *Report of a WHO informal consultation on SV40 and poliovaccines, Geneva, 18 September 1997.* Geneva, World Health Organization, 1997 (unpublished document; available on request from Quality Assurance and Safety: Biologicals, World Health Organization, 1211 Geneva 27, Switzerland).

6. *Polio: the beginning of the end.* Geneva, World Health Organization, 1997 (unpublished document, WHO/EPI/GEN/97.03; available on request from Expanded Programme on Immunization, World Health Organization, 1211 Geneva 27, Switzerland).
7. *Proposed global action plan and timetable for safe handling and maximum laboratory containment of wild polioviruses and potentially infectious materials: version for public comment, June 1998.* Geneva, World Health Organization, 1998 (unpublished document, WHO/EPI/GEN/98.05; available on request from Expanded Programme on Immunization, World Health Organization, 1211 Geneva 27, Switzerland).

WHO Technical Report Series, No. 897, 2000

Annex 3

Recommendations and guidelines for biological substances used in medicine and other documents[1]

The recommendations (previously called requirements) and guidelines published by the World Health Organization are scientific and advisory in nature but may be adopted by a national control authority as national requirements or used as the basis of such requirements.

These international recommendations are intended to provide guidance to those responsible for the production of biologicals as well as to others who may have to decide upon appropriate methods of assay and control in order to ensure that these products are safe, reliable and potent.

Recommendations concerned with biological substances used in medicine are formulated by international groups of experts and are published in the WHO Technical Report Series (TRS), as listed here. A historical list of requirements and other sets of recommendations is available on request from Quality Assurance and Safety: Biologicals, World Health Organization, 1211 Geneva 27, Switzerland.

Recommendations, guidelines and other documents

Recommendations and Guidelines	Reference
Acellular pertussis component of monovalent or combined vaccines	Adopted 1996, TRS 878 (1998)
Animal Cells, use of, as in vitro Substrates for the Production of Biologicals	Revised 1996, TRS 878 (1998)
Anthrax Spore Vaccine (Live, for Veterinary Use)	Adopted 1966, TRS 361 (1967)
BCG Vaccine, Dried	Revised 1985, TRS 745 (1987); Amendment 1987, TRS 771 (1988)
Biological products prepared by Recombinant DNA technology	Adopted 1990, TRS 814 (1991)
Brucella abortus Strain 19 Vaccine (Live, for Veterinary Use)	Adopted 1969, TRS 444 (1970); Addendum 1975, TRS 594 (1976)
Brucella melitensis Strain Rev. 1 Vaccine (Live, for Veterinary Use)	Adopted 1976, TRS 610 (1977)

[1] This Annex was updated during the preparation of the report for publication.

Recommendations and Guidelines	Reference
Collection, Processing and Quality Control of Blood, Blood Components and Plasma Derivatives	Revised 1992, TRS 840 (1994)
Diphtheria, Tetanus, Pertussis and Combined Vaccines	Revised 1989, TRS 800 (1990)
DNA Vaccines	Adopted 1996, TRS 878 (1998)
Haemophilus influenzae Type b Conjugate Vaccines	Revised 1998, TRS 897
Haemorrhagic Fever with Renal Syndrome (HFRS) Vaccine (Inactivated)	Adopted 1993, TRS 848 (1994)
Hepatitis B Vaccine prepared from Plasma	Revised 1987, TRS 771 (1988)
Hepatitis B Vaccines made by Recombinant DNA Techniques	Adopted 1988, TRS 786 (1989); Amendment 1997, TRS 889 (1999)
Human Interferons made by Recombinant DNA Techniques	Adopted 1987, TRS 771 (1988)
Human Interferons prepared from Lymphoblastoid Cells	Adopted 1988, TRS 786 (1989)
Immune Sera of Animal Origin	Adopted 1968, TRS 413 (1969)
Influenza Vaccine (Inactivated)	Revised 1990, TRS 814 (1991)
Influenza Vaccine (Live)	Adopted 1978, TRS 638 (1979)
Japanese Encephalitis Vaccine (Inactivated) for Human Use	Adopted 1987, TRS 771 (1988)
Louse-Borne Human Typhus Vaccine (Live)	Adopted 1982, TRS 687 (1983)
Measles, Mumps and Rubella Vaccines and Combined Vaccine (Live)	Adopted 1992, TRS 849 (1994); Note TRS 848 (1994)
Meningococcal Polysaccharide Vaccine	Adopted 1975, TRS 594 (1976); Addendum 1980, TRS 658 (1981)
Monoclonal Antibodies	Adopted 1991, TRS 822 (1992)
Poliomyelitis Vaccine (Inactivated)	Revised 1981, TRS 673 (1987); Addendum 1985, TRS 745 (1987)
Poliomyelitis Vaccine, Oral	Revised 1999, 50th report (in press); Addendum 1998, TRS 897
Rabies Vaccine (Inactivated) for Human Use, produced in Continuous Cell Lines	Adopted 1986, TRS 760 (1987); Amendment 1992, TRS 840 (1994)
Rabies Vaccine for Human Use	Revised 1980, TRS 658 (1981); Amendment 1992, TRS 840 (1994)
Rabies Vaccine for Veterinary Use	Adopted 1980, TRS 658 (1981); Amendment 1992, TRS 840 (1994)
Rift Valley Fever Vaccine	Adopted 1981, TRS 673 (1982)
Rift Valley Fever Vaccine (Live, Attenuated) for Veterinary Use	Adopted 1983, TRS 700 (1984)
Rinderpest Cell Culture Vaccine (Live) and Rinderpest Vaccine (Live)	Adopted 1969, TRS 444 (1970)
Smallpox Vaccine	Adopted 1966, TRS 323 (1966)
Snake Antivenins	Adopted 1970, TRS 463 (1971)

Recommendations and Guidelines	Reference
Sterility of Biological Substances	Revised 1973, TRS 530 (1973); Amendment 1995, TRS 872 (1998)
Synthetic Peptide Vaccines	Adopted 1997, TRS 889 (1999)
Thromboplastins and Plasma used to Control Oral Anticoagulant Therapy	Revised 1997, TRS 889 (1999)
Tick-borne Encephalitis Vaccine (Inactivated)	Adopted 1997, TRS 889 (1999)
Tuberculins	Revised 1985, TRS 745 (1987)
Typhoid Vaccine	Adopted 1966, TRS 361 (1967)
Typhoid Vaccine (Live, Attenuated) for Veterinary Use	Adopted 1983, TRS 700 (1984)
Varicella Vaccine (Live)	Revised 1993, TRS 848 (1994)
Vi Polysaccharide Typhoid Vaccine	Adopted 1992, TRS 840 (1994)
Yellow Fever Vaccine	Revised 1995, TRS 872 (1998)

Other documents	Reference
A review of tests on virus vaccines	TRS 673 (1982)
Biological standardization and control: a scientific review commissioned by the UK National Biological Standards Board (1997)	unpublished document WHO/BLG/97.1
Development of national assay services for hormones and other substances in community health care	TRS 565 (1975)
Good manufacturing practices for biological products	TRS 822 (1992)
Guidelines for national authorities on quality assurance for biological products	TRS 822 (1992)
Guidelines for quality assessment of antitumour antibiotics	TRS 658 (1981)
Laboratories approved by WHO for the production of yellow fever vaccine, revised 1995	TRS 872 (1998)
Procedure for approval by WHO of yellow fever vaccines in connexion with the issue of international vaccination certificates	TRS 658 (1981)
Production and testing of WHO yellow fever virus primary seed lot 213-77 and reference batch 168-73	TRS 745 (1987)
Recommendations for the assessment of binding-assay systems (including immunoassay and receptor assay systems) for human hormones and their binding proteins (A guide to the formulation of requirements for reagents and assay kits for the above assays and notes on cytochemical bioassay systems)	TRS 565 (1975)
Regulation and licensing of biological products in countries with newly developing regulatory authorities	TRS 858 (1987)

Other documents	Reference
Report of a WHO Consultation on Medicinal and other Products in Relation to Human and Animal Transmissible Spongiform Encephalopathies (1997)	unpublished documents WHO/EMC/Z00/97.3 WHO/BLG/97.2
Report of a WHO Meeting on Hepatitis B Vaccines Produced by Recombinant DNA Techniques	TRS 760 (1987)
Report on the standardization and calibration of cytokine immunoassays	TRS 889 (1999)
Standardization of interferons (reports of WHO informal consultations)	TRS 687 (1983) TRS 725 (1985) TRS 771 (1988)
Summary protocol for the batch release of virus vaccines	TRS 822 (1992)

WHO Technical Report Series, No. 897, 2000

Annex 4

International Biological Reference Preparations[1]

General information

The International Biological Reference Preparations provide a means of ensuring uniformity throughout the world in the designation of the potency or activity of biological substances used in the prophylaxis, therapy or diagnosis of disease, where this cannot be expressed in terms of physical or chemical quantities. For this purpose, International Units have been assigned, where necessary, to biological substances.

International Biological Reference Preparations, which are established by the WHO Expert Committee on Biological Standardization, have been denoted, variously, as International Reference Preparations, International Reference Reagents and International Standards. Biological reference preparations so established are by definition "primary reference materials". These preparations are intended for use in the calibration of the activity of secondary reference materials, either national or working standards, and for their expression in International Units. The World Health Assembly (resolution WHA 37.27) has recommended that Member States of the Organization give official recognition to existing International Standards.

The speed of development of new biological products, particularly cytokines and growth factors, revealed the need, often before the full programme for establishment of an International Standard could be completed, for the adoption of interim reference materials with an official status conferred by WHO. This need arose from a combination of regulatory and scientific considerations. Reference materials adopted

[1] This list was updated during preparation of the report for publication.

in this way are named interim Reference Reagents and are assigned a unitage expressed as "units".

The WHO Expert Committee on Biological Standardization (ECBS) meets annually and new additions and discontinuations of biological reference preparations are decided each year. The list of current reference materials, which incorporates the changes agreed at the Committee's forty-ninth meeting, is set out below. It replaces the list published in 1991.[1] During its preparation, changes agreed at the Committee's subsequent (fiftieth) meeting were also incorporated. An updated list of International Biological Reference Preparations will be available on the Internet at the following address: http://www.who.int/technology/biological.html.

Users will have two choices in searching for preparations: by alphabetical order (as in this Annex), or by class of substance. The lists are presented in tabular form and contain the following information:

- Preparation — Name of the preparation with pharmaceutical form and assigned unitage
- Standard — Last standard established and year of establishment
- WHO TRS ECBS report — Reference to the relevant ECBS report in the WHO Technical Report Series. Available from WHO[2]
- Material — Type/characteristics of the material in the preparation
- Held at — Custodian laboratory from which the preparation can be requested
- Code — Code for the preparation to be used when ordering the preparation
- BS document — Reports of WHO collaborative studies, data supporting the calibration of the preparation. Available from WHO[3]

Custodian laboratories

The main custodians of International Biological Reference Preparations are the International Laboratories for Biological Standards: the National Institute for Biological Standards and Control, Potters Bar, England, and the Central Laboratory of the Netherlands Red Cross Blood Transfusion Service, Amsterdam, Netherlands. Other custodian

[1] *Biological substances: International Standards and Reference Reagents 1990.* Geneva, World Health Organization, 1991.

[2] Marketing and Dissemination, World Health Organization, 1211 Geneva 27, Switzerland.

[3] Quality Assurance and Safety: Biologicals, Health Technology and Pharmaceuticals, World Health Organization, 1211 Geneva 27, Switzerland.

laboratories are the Anti-Viral Research Branch of the National Institute of Allergy and Infectious Diseases, Bethesda, MD, USA, and the Centers for Disease Control and Prevention, Atlanta, GA, USA.

These laboratories distribute samples of standards free of charge to national control laboratories for biological products and with small handling charges to other organizations, such as manufacturers and hospital laboratories. These preparations are generally intended for use in the calibration of the activity of secondary reference preparations (regional, national or in-house working standards) and, in most cases, for the expression of their biological activity in International Units. They are made for use in laboratory assays only and should not be administered to humans. Requests for International Reference Preparations should be addressed directly to the custodian laboratories, together with a statement of intended use. The addresses, telephone and fax numbers, and electronic mail (email) addresses of WHO custodian laboratories are as follows:

Centers for Disease Control and Prevention:

WHO Collaborating Centre for Reference and Research in Blood Lipids
Environmental Health Laboratory Sciences F25
National Center for Environmental Health
Centers for Disease Control and Prevention
Atlanta
GA 30341-3724
USA
Tel: +1 770 488 7952
Fax: +1 770 488 4839

Central Laboratory of the Netherlands Red Cross Blood Transfusion Service:

International Laboratory for Biological Standards
Central Laboratory of the Netherlands Red Cross Blood
Transfusion Service
125 Plesmanlaan
1066 CX Amsterdam
Post Box 9190
1066 AD Amsterdam
Netherlands
Tel: +31 20 512 9222
Fax: +31 20 512 3252

National Institute of Allergy and Infectious Diseases:

Division of Microbiology and Infectious Diseases
National Institute of Allergy and Infectious Diseases
National Institutes of Health
Solar Building – Room 3A22
6003 Executive Boulevard
Rockville
MD 20892
USA
Tel: +1 301 496 8285
Fax: +1 301 402 1456

National Institute for Biological Standards and Control:

International Laboratory for Biological Standards
National Institute for Biological Standards and Control
Blanche Lane
South Mimms
Potters Bar
Herts. EN6 3QH
England
Tel: +44 1707 646399
Fax: +44 1707 646977
Email: standards@nibsc.ac.uk

WHO International Biological Reference Preparations

Held and distributed by the WHO International Laboratories for Biological Standards

Preparation	Standard	WHO TRS ECBS Report	Material	Held at[a]	Code	WHO/BS Documents
Activin A, human, recombinant. Lyophilized. 5 units/ampoule.	1st Reference Reagent, 1998	No. 897, 49th Report	Recombinant hormone	NIBSC	91/626	98.1882
Alphafetoprotein, human. Lyophilized. 100 000 IU/ampoule.	1st International Standard, 1975	No. 594, 27th Report	Human cord serum	NIBSC	AFP	75.1121
Amikacin. Lyophilized. 50 600 IU/ampoule.	1st International Standard, 1983	No. 700, 34th Report	Antibiotic	NIBSC	80/541	83.1398
Amphotericin B. Lyophilized. 940 IU/mg. Approximately 50 mg of Amphotericin B.	1st International Standard, 1963	No. 274, 16th Report	Antibiotic	NIBSC	60/26	63.648
Ancrod. Lyophilized. 55 IU/ampoule.	1st International Reference Preparation, 1976	No. 626, 29th Report	Enzyme	NIBSC	74/581	76.1130
Anthrax spore vaccine. Lyophilized spore suspension of *Bacillus anthracis* strain 34 F2. 1 IU/ampoule.	1st International Reference Preparation, 1978	No. 638, 30th Report	Antigen	NIBSC	AxV2	78.1198
Anti-A blood-typing serum, human. Lyophilized. 470 IU/ampoule.	2nd International Standard, 1981	No. 673, 32nd Report	Human serum	CLB	W1001	81.1309
Anti-B blood-typing serum, human. Lyophilized. 860 IU/ampoule.	3rd International Standard, 1981	No. 673, 32nd Report	Human serum	CLB	W1002	81.1309
Anti-*Brucella abortus* serum, bovine. Lyophilized. 1000 IU of agglutinating activity and 1000 IU of complement-fixing activity per ampoule.	2nd International Standard, 1975	No. 594, 27th Report	Bovine serum	NIBSC	BaDS	75.1124
Anti-*Brucella ovis* serum, ovine. Lyophilized. 1000 IU/ampoule.	1st International Standard, 1985	No. 745, 36th Report	Ovine serum	NIBSC	BovisDs	85.1485
Anti-C complete blood-typing serum, human. Lyophilized. 100 IU/ampoule.	1st International Standard, 1984	No. 725, 35th Report	Human serum	CLB	W1004	84.1424

Preparation	Standard	WHO TRS ECBS Report	Material	Held at[a]	Code	WHO/BS Documents
Anti-C incomplete blood-typing serum, human. Lyophilized. 64 IU/ampoule.	1st International Standard, 1976	No. 610, 28th Report	Human serum	CLB	W1007	71.1038, 76.1130
Anti-canine distemper serum. Lyophilized. 1000 IU/ampoule.	1st International Standard, 1967	No. 384, 20th Report	Horse serum	NIBSC	CDS	67.881, 67.881 Add. 1
Anti-canine hepatitis serum. Lyophilized. 1000 IU/ampoule.	1st International Standard, 1967	No. 384, 20th Report	Horse serum	NIBSC	CHS	67.880, 67.880 Add. 1
Anti-D (anti-RhO) complete blood-typing serum, human. Lyophilized. 128 IU/ampoule.	1st International Standard, 1990	No. 814, 41st Report	Human serum, chemically modified	CLB	64/16	90.1645
Anti-D (anti-RhO) incomplete blood-typing serum, human. Lyophilized. 32 IU/ampoule.	1st International Standard, 1966	No. 361, 19th Report	Human serum	CLB	W1006	66.810
Anti-D immunoglobulin, human. Lyophilized. 300 IU/ampoule (60 μg/ampoule).	1st International Reference Preparation, 1977	No. 626, 29th Report	Human immunoglobulin	NIBSC	68/419	76.1130, 78.1236
Anti-double-stranded DNA serum, human. Lyophilized. 100 IU/ampoule.	1st International Standard, 1985	No. 745, 36th Report	Human serum	CLB	Wo/80	85.1491
Anti-dysentery serum (Shiga), equine. 1 IU equivalent to 0.05 mg of dried standard serum.	1st International Standard, 1928	—	Horse serum	NIBSC	DY	*Bull. Health Org.*, 4, 1935
Anti-E complete blood-typing serum, human. Lyophilized. 100 IU/ampoule.	1st International Standard, 1983	No. 700, 34th Report	Human serum	CLB	W1005	83.1424
Anti-*Echinococcus* serum, human. Lyophilized. 87.36 mg human serum per ampoule.	1st International Reference Reagent, 1975	No. 594, 27th Report	Human serum	NIBSC	ECHS	75.1106
Anti-hepatitis A immunoglobulin, human. Lyophilized. 49 IU/ampoule.	2nd International Standard, 1998	No. 897, 49th Report	Human immunoglobulin	CLB	97/646	98.1878, 98.1878 Add. 1
Anti-hepatitis B immunoglobulin, human. Lyophilized. 50 IU/ampoule.	1st International Reference Preparation, 1977	No. 626, 29th Report	Human immunoglobulin	CLB	W1042	77.1164

Preparation	Standard	WHO TRS ECBS Report	Material	Held at[a]	Code	WHO/BS Documents
Anti-hepatitis E serum, human. Lyophilized. 50 units/ampoule.	1st Reference Reagent, 1997	No. 889, 48th Report	Human serum	NIBSC	95/584	97.1869
Anti-human platelet antigen-1a. Lyophilized. No unitage assigned.	1st Reference Reagent, 1997	No. 889, 48th Report	Human serum	NIBSC	93/710	97.1862
Anti-interferon alpha serum, human. Lyophilized. 8000 neutralizing units/ampoule.	1st International Reference Reagent, 1994	No. 858, 45th Report	Human serum	NIAID	G037-501-572	94.1783
Anti-interferon beta serum, human. Lyophilized. 1500 neutralizing units/ampoule.	1st International Reference Reagent, 1994	No. 858, 45th Report	Human serum	NIAID	G038-501-572	94.1783
Anti-measles serum, human. Lyophilized. 5 IU/ampoule.	2nd International Standard, 1990	No. 814, 41st Report	Human serum	NIBSC	66/202	90.1636
Anti-*Mycoplasma gallisepticum* serum. Lyophilized. 1000 IU of agglutinating activity and 1000 IU of haemagglutination-inhibiting activity per ampoule.	1st International Reference Preparation, 1969	No. 444, 22nd Report	Chicken serum	NIBSC	MgDS	69.961
Anti-Newcastle-disease serum. Lyophilized. 320 IU/ampoule.	1st International Reference Preparation, 1968	No. 413, 21st Report	Chicken serum	NIBSC	NDS	67.878, 68.935
Anti-nuclear ribonucleoprotein serum. Lyophilized. No unitage assigned.	1st International Reference Reagent, 1983	No.700, 34th Report	Human serum	CLB	W1063	83.1390
Anti-nuclear-factor (homogeneous) serum. Lyophilized. No unitage assigned.	1st International Reference Preparation, 1970	No. 463, 23rd Report	Human serum	CLB	66/233	70.1005
Anti-parvovirus B19 (IgG) serum. Lyophilized. 100 IU/ampoule.	1st International Standard, 1995	No. 872, 46th Report	Human serum	NIBSC	93/724	95.1810
Anti-pertussis serum, mouse. 17 units of anti-pertussis toxin, 143 units of anti-filamentous haemagglutinin, 30 units of anti-pertactin and 32 units of anti-fimbria types 2/3 per ampoule.	1st Reference Reagent, 1999	50th Report (in press)	Mouse serum	NIBSC	97/642	99.1901

Preparation	Standard	WHO TRS ECBS Report	Material	Held at[a]	Code	WHO/BS Documents
Anti-poliovirus serum (types 1, 2, 3). Lyophilized. 25 IU/ampoule (type 1), 50 IU/ampoule (type 2), 5 IU/ampoule (type 3).	2nd International Standard, 1991	No. 822, 42nd Report	Human serum	NIBSC	66/202	91.1660
Anti-Q-fever serum, bovine. Lyophilized. 1000 IU/ampoule.	1st International Standard, 1954	No. 96, 8th Report	Bovine serum	NIBSC	QF	53.230, 54.276
Anti-rabies immunoglobulin, human. Lyophilized. 30 IU/ampoule.	2nd International Standard, 1993	No. 848, 44th Report	Human immunoglobulin	NIBSC	RAI	93.1749, 93.1749 Add. 1
Anti-rubella immunoglobulin, human. Lyophilized. 1600 IU/vial.	1st International Standard, 1996	No. 878, 47th Report	Human immunoglobulin	NIBSC	RUBI-1-94	96.1833
Anti-*Salmonella pullorum* serum (standard form S). Lyophilized. 1000 IU/ampoule.	1st International Standard, 1973	No. 530, 25th Report	Goat serum	NIBSC	SpDS-S2	73.1066
Anti-*Salmonella pullorum* serum (variant form V). Lyophilized. 1000 IU/ampoule.	1st International Standard, 1973	No. 530, 25th Report	Goat serum	NIBSC	SpDS-V	73.1066, 73.1066 Corr. 1
Anti-smooth muscle serum. Lyophilized. No unitage assigned.	1st International Reference Reagent, 1983	No. 700, 34th Report	Human serum	CLB	W1062	83.1390
Anti-staphylococcal P-V leucocidin serum, equine. Lyophilized. 150 IU/ampoule.	1st International Standard, 1965	No. 329, 18th Report	Horse serum	NIBSC	SPLS	65.769, 66.816
Anti-swine fever serum. Lyophilized. 1000 IU/ampoule.	1st International Standard, 1963	No. 274, 16th Report	Pig serum	NIBSC	SFS	62.573, 63.627, 63.627 Add. 1
Anti-syphilitic serum, human. Lyophilized. 49 IU/ampoule.	1st International Standard, 1958	No. 172, 12th Report	Human serum	NIBSC	HS	58.439
Anti-tetanus immunoglobulin, human. Lyophilized. 120 IU/ampoule.	1st International Standard, 1992	No. 840, 43rd Report	Human immunoglobulin	NIBSC	TE-3	92.1696, 92.1696 Add. 1

Preparation	Standard	WHO TRS ECBS Report	Material	Held at[a]	Code	WHO/BS Documents
Anti-tetanus serum, equine. Lyophilized. 1400 IU/ampoule.	2nd International Standard, 1969	No. 444, 22nd Report	Horse serum	NIBSC	TE	69.964
Anti-thyroglobulin serum, human. Lyophilized. 1000 IU/ampoule.	1st International Reference Preparation, 1978	No. 638, 30th Report	Human serum	NIBSC	65/93	78.1188, 78.1189
Anti-tick-borne encephalitis serum (louping ill (Moredun) virus). Lyophilized. No unitage assigned.	1st International Reference Reagent, 1964	No. 293, 17th Report	Sheep serum	NIBSC	TILI	64.707, 65.757, 65.757 Corr. 1
Anti-tick-borne encephalitis serum (Russian spring–summer (Sophyn and Absettarov) virus). Lyophilized. No unitage assigned.	1st International Reference Reagent, 1964	No. 293, 17th Report	Sheep serum	NIBSC	TISA	64.707, 65.757, 65.757 Corr. 1
Anti-toxoplasma serum. Lyophilized. 1000 IU/ampoule.	3rd International Standard, 1994	No. 858, 45th Report	Human serum	NIBSC	TOXM	94.1761
Anti-typhoid serum, equine. Lyophilized. 5 ml of freeze-dried horse serum per ampoule.	1st International Reference Preparation, 1957	No. 68, 6th Report	Horse serum	NIBSC	TYS	57.415
Anti-varicella zoster immunoglobulin, human. Lyophilized. 50 IU/ampoule.	1st International Standard, 1987	No. 771, 38th Report	Human immunoglobulin	CLB	W1044	87.1565
Anti-yellow fever serum, monkey. Lyophilized. 143 IU/ampoule.	1st International Standard, 1962	No. 259, 15th Report	Monkey serum	NIBSC	YF	62.545, 62.545 Corr. 1
Antithrombin, concentrate, human. Lyophilized. Functional potency 4.7 IU/ampoule; antigenic potency 5.1 IU/ampoule.	2nd International Standard, 1997	No. 889, 48th Report	Purified plasma protein	NIBSC	96/520	97.1863
Antithrombin, plasma, human. Lyophilized. 0.85 IU/ampoule.	2nd International Standard, 1994	No. 848, 44th Report	Human plasma	NIBSC	93/768	94.1785
Apolipoprotein A-1. Lyophilized. 1.66 mg/vial.	1st International Reference Reagent, 1992	No. 840, 43rd Report	Human serum	CDC	SP1-01	92.1706

Preparation	Standard	WHO TRS ECBS Report	Material	Held at[a]	Code	WHO/BS Documents
Apolipoprotein B. Frozen. 1.22 g/l. Vials, 1 ml.	1st International Reference Reagent, 1993	No. 848, 44th Report	Human serum	CDC	SP3-07	93.1721
Arginine vasopressin. Lyophilized. 8.2 IU/ampoule.	1st International Standard, 1978	No. 638, 30th Report	Hormone	NIBSC	77/501	78.1231
Atrial natriuretic factor, human. Lyophilized. 2.5 IU/ampoule.	1st International Standard, 1987	No. 771, 38th Report	Peptide hormone	NIBSC	85/669	87.1578
Bacitracin. Lyophilized. Approximately 100 mg of zinc bacitracin (74 IU/mg).	2nd International Standard, 1964	No. 293, 17th Report	Antibiotic	NIBSC	62/3	64.681
Basic fibroblast growth factor (bFGF, FGF-2). Lyophilized. 1600 IU/ampoule.	1st International Standard, 1993	No. 848, 44th Report	Recombinant growth factor	NIBSC	90/712	93.1736
BCG vaccine. Lyophilized. 2.5 mg of semi-dry bacillary mass per 5-ml ampoule.	1st International Reference Preparation, 1965	No. 329, 18th Report	Antigen	NIBSC	BCG	65.802, 82.1387, 90.1638
Beta-2 microglobulin. Lyophilized. 100 IU/ampoule.	1st International Standard, 1985	No. 745, 36th Report	Human serum	NIBSC	B2M	85.1501
Beta-thromboglobulin, human. Lyophilized. 500 IU/ampoule.	1st International Standard, 1984	No. 725, 35th Report	Purified plasma protein	NIBSC	83/501	84.1455
Birch pollen extract (*Betula verrucosa*). Lyophilized. 100 000 IU/ampoule.	1st International Standard, 1986	No. 760, 37th Report	Allergen	NIBSC	84/522	86.1512
Bleomycin complex A2/B2. Lyophilized. 8910 IU/ampoule.	1st International Reference Preparation, 1980	No. 658, 31st Report	Antibiotic	NIBSC	78/547	80.1276, 80.1276 Add. 1
Blood coagulation factor VII concentrate, human. Lyophilized. 6.3 IU/ampoule.	1st International Standard, 1998	No. 897, 49th Report	Purified plasma protein	NIBSC	97/592	98.1876
Blood coagulation factor VIIa concentrate, human. Lyophilized. 5130 IU/ampoule.	1st International Standard, 1993	No. 848, 44th Report	Recombinant protein	NIBSC	89/688	93.1728
Blood coagulation factor VIII concentrate, human. Lyophilized. 8.5 IU/ampoule.	6th International Standard, 1998	No. 897, 49th Report	Recombinant protein	NIBSC	97/616	98.1888, 98.1888 Add. 1

Preparation	Standard	WHO TRS ECBS Report	Material	Held at[a]	Code	WHO/BS Documents
Blood coagulation factor IX concentrate, human. Lyophilized. 10.7 IU/ampoule.	3rd International Standard, 1996	No. 878, 47th Report	Purified plasma protein	NIBSC	96/854	96.1845
Blood coagulation factor IXa, concentrate, human. Lyophilized. 11.0 IU/ampoule.	1st International Standard, 1999	50th Report (in press)	Recombinant protein	NIBSC	97/562	99.1916
Blood coagulation factors II and X concentrate, human. Lyophilized. 11.2 IU of factor II and 10.2 IU of factor X per ampoule.	3rd International Standard, 1999	50th Report (in press)	Purified plasma proteins	NIBSC	98/590	99.1904
Blood coagulation factor VIII and von Willebrand factor, plasma, human. Lyophilized. 0.57 IU of factor VIII clotting activity, 0.89 IU of factor VIII antigen, 0.73 IU of von Willebrand antigen and 0.73 IU of von Willebrand ristocetin cofactor activity per ampoule.	4th International Standard, 1998	No. 897, 49th Report	Human plasma	NIBSC	97/586	98.1881
Blood coagulation factors II, VII, IX, X, plasma, human. Lyophilized. 0.93 IU of factor II, 1.25 IU of factor VII, 0.90 IU of factor IX and 0.95 IU of factor X per ampoule.	2nd International Standard, 1996	No. 878, 47th Report	Human plasma	NIBSC	94/746	96.1840
Bone morphogenic protein-2. Lyophilized. 5000 units/ampoule.	1st Reference Reagent, 1997	No. 889, 48th Report	Recombinant cytokine	NIBSC	93/574	97.1857
Brain-derived neurotrophic factor. Lyophilized. 16 000 units/ampoule.	1st Reference Reagent, 1997	No. 889, 48th Report	Recombinant growth factor	NIBSC	96/534	97.1867
C-reactive protein, human. Lyophilized. 0.049 IU/ampoule.	1st International Standard, 1986	No. 760, 37th Report	Human plasma	NIBSC	85/506	86.1514
Calcitonin, eel. Lyophilized. 88 IU/ampoule.	1st International Standard, 1989	No. 800, 40th Report	Hormone	NIBSC	88/556	89.1620
Calcitonin, human. Lyophilized. 17.5 IU/ampoule.	2nd International Standard, 1991	No. 822, 42nd Report	Hormone	NIBSC	89/620	91.1675
Calcitonin, porcine. Lyophilized. 0.8 IU/ampoule.	2nd International Standard, 1991	No. 822, 42nd Report	Hormone	NIBSC	89/540	91.1674
Calcitonin, salmon. Lyophilized. 138 IU/ampoule.	3rd International Standard, 1999	50th Report (in press)	Hormone	NIBSC	98/586	99.1906, 99.1906 Add. 1

Preparation	Standard	WHO TRS ECBS Report	Material	Held at[a]	Code	WHO/BS Documents
Capreomycin. Lyophilized. 920 IU/mg. Approximately 80 mg of capreomycin sulfate.	1st International Reference Preparation, 1967	No. 384, 20th Report	Antibiotic	NIBSC	67/150	67.884
Carcinoembryonic antigen (CEA), human. Lyophilized. 100 IU/ampoule.	1st International Reference Preparation, 1975	No. 594, 27th Report	Derived from human carcinoma	NIBSC	73/601	75.1110, 78.1200
Chlortetracycline. Lyophilized. 1000 IU/mg. Approximately 75 mg of chlortetracycline hydrochloride.	2nd International Standard, 1969	No. 444, 22nd Report	Antibiotic	NIBSC	66/232	69.983
Cholera antitoxin, goat. Lyophilized. 2200 IU/ampoule.	1st International Standard, 1984	No. 725, 35th Report	Goat serum	NIBSC	CHAN	84.1438
Cholera vaccine (Inaba). Lyophilized. 40 000 000 000 organisms/ampoule.	2nd International Reference Preparation, 1971	No. 486, 24th Report	Antigen	NIBSC	INV	71.1032 Rev. 1
Cholera vaccine (Ogawa). Lyophilized. 40 000 000 000 organisms/ampoule.	2nd International Reference Preparation, 1971	No. 486, 24th Report	Antigen	NIBSC	OGV	71.965
Chorionic gonadotrophin, alpha subunit, human, for immunoassay. Lyophilized. 70 IU/ampoule.	1st International Reference Preparation, 1974	No. 565, 26th Report	Glycoprotein hormone	NIBSC	75/569	74.1094
Chorionic gonadotrophin, beta subunit, human, for immunoassay. Lyophilized. 70 IU/ampoule.	1st International Reference Preparation, 1974	No. 565, 26th Report	Glycoprotein hormone	NIBSC	75/551	74.1094
Chorionic gonadotrophin, human. Lyophilized. 650 IU/ampoule.	4th International Standard, 1999	50th Report (in press)	Glycoprotein hormone	NIBSC	75/589	99.1905
***Clostridium botulinum* Type B antitoxin, equine**. Lyophilized. 30.7 IU/ampoule.	2nd International Standard, 1985	No. 745, 36th Report	Horse serum	NIBSC	BUSB	85.1469
***Clostridium botulinum* Type E antitoxin, equine**. Lyophilized. 1000 IU/ampoule.	1st International Standard, 1962	No. 259, 15th Report	Horse serum	NIBSC	BTUSE	62.582

Preparation	Standard	WHO TRS ECBS Report	Material	Held at[a]	Code	WHO/BS Documents
***Clostridium novyi* alpha toxoid**. Lyophilized. No unitage assigned.	1st International Reference Preparation, 1966	No. 361, 19th Report	Toxoid	NIBSC	CoT	70.1022, 66.819
***Clostridium perfringens* beta antitoxin, equine**. Lyophilized. 4770 IU/ampoule.	2nd International Standard, 1998	No. 897, 49th Report	Horse serum	NIBSC	2Cp BetaAt	85.1484
***Clostridium perfringens* beta toxoid**. Lyophilized. No unitage assigned.	1st International Reference Preparation, 1975	No. 594, 27th Report	Toxoid	NIBSC	CWBetatd	75.1122
***Clostridium perfringens* epsilon antitoxin, equine**. Lyophilized. 1020 IU/ampoule.	2nd International Standard, 1985	No. 745, 36th Report	Horse serum	NIBSC	2Cp Epsilon At	85.1484
***Clostridium perfringens* epsilon toxoid**. Lyophilized. No unitage assigned.	1st International Reference Preparation, 1975	No. 594, 27th Report	Toxoid	NIBSC	CW Epsilon td	75.1122
Colistin. Lyophilized. 20 500 IU/mg. Approximately 75 mg of colistin sulfate.	1st International Standard, 1968	No. 413, 21st Report	Antibiotic	NIBSC	65/62	68.923
Colistin methane sulfonate. Lyophilized. 12 700 IU/mg. Approximately 75 mg of colistin methane sulfonate.	1st International Reference Preparation, 1968	No. 413, 21st Report	Antibiotic	NIBSC	66/254	68.924
Corticotrophin (ACTH), porcine. Lyophilized. 5 IU/ampoule.	3rd International Standard, 1962	No. 259, 15th Report	Pituitary hormone	NIBSC	59/16	62.548
Dihydrostreptomycin. Lyophilized. 820 IU/mg. Approximately 200 mg of dihydrostreptomycin sulfate.	2nd International Standard, 1966	No. 361, 19th Report	Antibiotic	NIBSC	62/13	66.829
Diphtheria (Schick) test toxin. Lyophilized. 900 IU/ampoule.	1st International Standard, 1954	No. 96, 8th Report	Toxin	NIBSC	STT	54.274, 54.275, 54.275 Add. 1, Add. 2
Diphtheria antitoxin, equine. Lyophilized. Bottles in the form of 10 ml of a solution of dried serum in 66% v/v of glycerol. 10 IU/ml.	1st International Standard, 1934	—	Horse serum	NIBSC	97/762	*Bull. Health Org.* 5, 1936

Preparation	Standard	WHO TRS ECBS Report	Material	Held at[a]	Code	WHO/BS Documents
Diphtheria toxoid, adsorbed. Lyophilized. 160 IU/ampoule.	3rd International Standard, 1999	50th Report (in press)	Adsorbed toxoid	NIBSC	98/650	99.1915, 99.1915 Add. 1
Diphtheria toxoid, for flocculation tests. Lyophilized. 900 Lf IU/ampoule.	1st International Reference Reagent, 1988	No. 786, 39th Report	Toxoid	NIBSC	DIFT	88.1590
Dog hair and dander extract (*Canis domesticus*). Lyophilized. 100 000 IU/ampoule.	1st International Standard, 1986	No. 760, 37th Report	Allergen	NIBSC	84/685	86.1513, 87.1547
Elcatonin. Lyophilized. 15 IU/ampoule.	1st International Standard, 1986	No. 760, 37th Report	Endocrinological products	NIBSC	84/614	85.1494, 86.1529
Endotoxin for *Limulus* amoebocyte lysate (LAL) gelation tests. Lyophilized. 10 000 IU/ampoule.	2nd International Standard, 1996	No. 878, 47th Report	Toxin	NIBSC	94/580	96.1830 Rev. 1
Epidermal growth factor (1-52), human, recombinant. Lyophilized. 1.75 µg/ampoule. No unitage assigned.	1st International Reference Reagent, 1994	No. 858, 45th Report	Recombinant growth factor	NIBSC	91/550	94.1781
Epidermal growth factor, human, recombinant. Lyophilized. 2000 IU/ampoule.	1st International Standard, 1994	No. 858, 45th Report	Recombinant growth factor	NIBSC	91/530	94.1781
Erythromycin. Lyophilized. 920 IU/mg. Approximately 75 mg of erythromycin A base.	2nd International Standard, 1978	No. 638, 30th Report	Antibiotic	NIBSC	76/538	78.1228
Erythropoietin, human, urinary. Lyophilized. 10 IU/ampoule.	2nd International Reference Preparation, 1970	No. 463, 23rd Report	Urinary hormone	NIBSC	67/343	70.1015
Erythropoietin, rDNA-derived. Lyophilized. 86 IU/ampoule.	1st International Standard, 1990	No. 814, 41st Report	Recombinant hormone	NIBSC	87/684	90.1650
Ferritin, human, recombinant. Lyophilized. 6.3 µg/ampoule.	3rd International Standard, 1996	No. 878, 47th Report	Recombinant protein diluted in plasma	NIBSC	94/572	96.1838, 91.1678
Fibrinogen, plasma, human. Lyophilized. 2.2 mg/ampoule.	2nd International Standard, 1999	50th Report (in press)	Human plasma	NIBSC	98/612	99.1903

Preparation	Standard	WHO TRS ECBS Report	Material	Held at[a]	Code	WHO/BS Documents
FMS-like tyrosine kinase 3 ligand. Lyophilized. 1000 units/ampoule.	1st Reference Reagent, 1997	No. 889, 48th Report	Recombinant cytokine	NIBSC	96/532	97.1858
Follicle-stimulating hormone, human, pituitary. Lyophilized. 80 IU/ampoule.	1st International Standard, 1986	No. 760, 37th Report	Pituitary hormone	NIBSC	83/575	86.1535
Follicle-stimulating hormone, human, recombinant, for bioassay. Lyophilized. 138 IU/ampoule.	1st International Standard, 1995	No. 872, 46th Report	Recombinant hormone	NIBSC	92/642	95.1819
Follicle-stimulating hormone, human, recombinant, for immunoassay. Lyophilized. 60 IU/ampoule.	1st International Standard, 1997	No. 889, 48th Report	Recombinant hormone	NIBSC	92/510	97.1871
Follicle-stimulating hormone, human, urinary (urofollitropin), for bioassay. Lyophilized. 121 IU/ampoule.	1st International Standard, 1995	No. 872, 46th Report	Urinary hormone	NIBSC	92/512	95.1819
Follicle-stimulating hormone and luteinizing hormone, human, urinary, for bioassay. Lyophilized. 54 IU of FSH and 46 IU of LH per ampoule.	3rd International Standard, 1993	No. 848, 44th Report	Urinary hormone	NIBSC	71/264	93.1743
Gas-gangrene antitoxin (*Clostridium histolyticum*), equine. Lyophilized. 50 IU/ampoule.	3rd International Standard, 1971	No. 486, 24th Report	Horse serum	NIBSC	HI	71.1034
Gas-gangrene antitoxin (*Clostridium novyi*), equine. Lyophilized. 1100 IU/ampoule.	3rd International Standard, 1966	No. 361, 19th Report	Horse serum	NIBSC	OE	66.803
Gas-gangrene antitoxin (*Clostridium perfringens*), equine. Lyophilized. 270 IU/ampoule.	5th International Standard, 1962	No. 259, 15th Report	Horse serum	NIBSC	PE	62.547 Rev. 1
Gas-gangrene antitoxin (*Clostridium septicum*), equine. Lyophilized. 500 IU/ampoule.	3rd International Standard, 1957	No. 147, 11th Report	Horse serum	NIBSC	VI	57.384
Gas-gangrene antitoxin (*Clostridium sordellii*), equine. Lyophilized. 7500 IU/ampoule. Bottles in the form of 10 ml of a solution of dried serum in 66% v/v of glycerol; 20 IU/ml.	1st International Standard, 1938	—	Horse serum	NIBSC	97/764	*Bull. Health Org.* 8, 1939
Gentamicin. Lyophilized. 31 020 IU/ampoule.	2nd International Standard, 1995	No. 872, 46th Report	Antibiotic	NIBSC	92/670	95.1811
Glucagon, porcine. Lyophilized. 1.49 IU/ampoule.	1st International Standard, 1973	No. 530, 25th Report	Gastrointestinal peptide	NIBSC	69/194	73.1064, 74.1092

Preparation	Standard	WHO TRS ECBS Report	Material	Held at[a]	Code	WHO/BS Documents
Gonadotrophin, equine serum. Lyophilized. 1600 IU/ ampoule.	2nd International Standard, 1966	No. 361, 19th Report	Horse serum	NIBSC	62/1	66.855
Gramicidin. Lyophilized. 1000 IU/mg. Approximately 55 mg of gramicidin.	1st International Reference Preparation, 1966	No. 361, 19th Report	Antibiotic	NIBSC	64/10	66.858
Granulocyte colony-stimulating factor, human, recombinant. Lyophilized. 10 000 IU/ampoule.	1st International Standard, 1992	No. 840, 43rd Report	Recombinant growth factor	NIBSC	88/502	92.1711
Granulocyte-macrophage colony-stimulating factor, human, recombinant. Lyophilized. 10 000 IU/ampoule.	1st International Standard, 1992	No. 840, 43rd Report	Recombinant growth factor	NIBSC	88/646	92.1711
Haemoglobin A2, raised. Lyophilized. 5.3 % (w/w) of total haemoglobin.	1st International Reference Reagent, 1993	No. 848, 44th Report	Derived from beta-thalassaemia haemolysate	NIBSC	89/666	93.1754
Haemoglobin F, raised. Lyophilized. 3.4% (w/w) of total haemoglobin.	1st International Reference Reagent, 1993	No. 848, 44th Report	Derived from cyanmethaemo-globin haemolysate	NIBSC	85/616	93.1755
Haemoglobincyanide. Liquid. 49.79 μmol/l.	6th International Standard, 1998	No. 897, 49th Report	Bovine haemoglobin	NIBSC	98/708	98.1886
Heparin, low molecular weight. Lyophilized. 1680 IU of anti-Xa activity and 665 IU of anti-IIa activity per ampoule.	1st International Standard, 1986	No. 760, 37th Report	Derived from porcine mucosa	NIBSC	85/600	86.1541
Heparin, low molecular weight (calibrant for molecular weight distribution). Lyophilized. No unitage assigned.	1st International Reference Reagent, 1993	No. 848, 44th Report	Derived from porcine mucosa	NIBSC	90/686	93.1727
Heparin, unfractionated. Lyophilized. 2031 IU/ampoule.	5th International Standard, 1998	No. 897, 49th Report	Derived from porcine mucosa	NIBSC	97/578	98.1875
Hepatitis A vaccine, inactivated. Frozen. Immunogenic activity 100 IU/ml; antigen content 100 IU/ml.	1st International Standard, 1999	50th Report (in press)	Antigen	NIBSC	95/500	99.1914
Hepatitis B surface antigen, ad subtype. Lyophilized. 100 IU/ampoule.	1st International Standard, 1985	No. 745, 36th Report	Human serum	NIBSC	80/549	85.1476 Rev. 1

Preparation	Standard	WHO TRS ECBS Report	Material	Held at[a]	Code	WHO/BS Documents
Hepatitis B vaccine, plasma derived, for immunogenicity studies. Liquid. 1.2 ml/ampoule. 20 μg hepatitis B surface antigen/ml.	1st International Reference Reagent, 1986	No. 760, 37th Report	Adsorbed antigen	NIBSC	85/65	86.1525 Rev. 1
Hepatitis B virus DNA. Lyophilized. 500 000 IU/vial.	1st International Standard, 1999	50th Report (in press)	Nucleic acid preparation	NIBSC	97/746	99.1917
Hepatitis C virus RNA. Lyophilized. 50 000 IU/vial.	1st International Standard, 1997	No. 889, 48th Report	Nucleic acid preparation	NIBSC	96/790	97.1861
Hepatocyte growth factor/scatter factor (precursor), human. Lyophilized. 2000 IU/ampoule.	1st International Standard, 1999	50th Report (in press)	Recombinant growth factor	NIBSC	96/556	99.1899
Hepatocyte growth factor/scatter factor, human. Lyophilized. 4000 IU/ampoule.	1st International Standard, 1999	50th Report (in press)	Recombinant growth factor	NIBSC	96/564	99.1899
HIV-1 p24 antigen. Lyophilized. 1000 IU/ampoule.	1st International Reference Reagent, 1992	No. 840, 43rd Report	Peptide in human serum	NIBSC	90/636	92.1699
HIV-1 RNA. Lyophilized. 100 000 IU/vial.	1st International Standard, 1999	50th Report (in press)	Nucleic acid preparation	NIBSC	97/656	99.1910
Horseradish peroxidase-conjugated sheep anti-human IgG (H and L chains). Lyophilized. No unitage assigned.	1st International Reference Preparation, 1982	No. 687, 33rd Report	Sheep serum	CLB	W1024	81.1342
House-dust mite extract (*Dermatophagoides pteronyssinus*). Lyophilized. 10 000 IU/ampoule.	1st International Standard, 1983	No. 700, 34th Report	Allergen	NIBSC	82/518	83.1417
Human growth hormone, pituitary. Lyophilized. 4.4 IU/ampoule.	1st International Standard, 1982	No. 687, 33rd Report	Pituitary hormone	NIBSC	80/505	82.1369
Human serum complement components C1q, C4, C5, factor B, and whole functional complement CH50. Lyophilized. 100 IU of each component per ampoule.	1st International Reference Preparation, 1980	No. 658, 31st Report	Human serum	CLB	W1032	80.1281
Human serum immunoglobulin E. Lyophilized. 5000 IU/ampoule.	2nd International Reference Preparation, 1980	No. 658, 31st Report	Human serum	NIBSC	75/502	79.1240, 70.1019

Preparation	Standard	WHO TRS ECBS Report	Material	Held at[a]	Code	WHO/BS Documents
Human serum immunoglobulins G, A, and M (IgG, IgA and IgM). Lyophilized. 100 IU of IgG, 100 IU of IgA and 100 IU of IgM per ampoule.	1st International Reference Preparation, 1970	No. 463, 23rd Report	Human serum	NIBSC	67/86	70.1019
Human serum proteins, for immunoassay: albumin; alpha-1-antitrypsin; alpha-2-macroglobulin; ceruloplasmin; complement C3; transferrin. Lyophilized. 100 IU of each protein per ampoule.	1st International Reference Preparation, 1977	No. 626, 29th Report	Human serum	CLB	W1031	77.1155
Inhibin, human, recombinant. Lyophilized. 150 000 IU/ampoule.	1st International Standard, 1994	No. 858, 45th Report	Recombinant hormone	NIBSC	91/624	94.1787
Inhibin, porcine. Lyophilized. 2000 IU/ampoule.	1st International Standard, 1990	No. 814, 41st Report	Purified porcine hormone	NIBSC	86/690	90.1648
Insulin C-peptide, human. Lyophilized. 10 μg/ampoule.	1st International Reference Reagent, 1986	No. 760, 37th Report	Synthetic C-peptide	NIBSC	84/510	86.1538
Insulin, bovine. Hydrated crystals. 25.7 IU/mg. Approximately 50 mg/ampoule.	1st International Standard, 1986	No. 760, 37th Report	Insulin crystals	NIBSC	83/511	86.1524
Insulin, human. Hydrated crystals. 26 IU/mg. Approximately 50 mg/ampoule.	1st International Standard, 1986	No. 760, 37th Report	Insulin crystals	NIBSC	83/500	86.1524
Insulin, human, for immunoassay. Lyophilized. 3 IU/ampoule.	1st International Reference Preparation, 1974	No. 565, 26th Report	Human insulin	NIBSC	66/304	74.1084
Insulin, porcine. Hydrated crystals. 26 IU/mg. Approximately 50 mg/ampoule.	1st International Standard, 1986	No. 760, 37th Report	Insulin crystals	NIBSC	83/515	86.1524
Insulin-like growth factor I, human, recombinant. Lyophilized. 150 IU/ampoule.	1st International Standard, 1994	No. 858, 45th Report	Recombinant growth factor	NIBSC	91/554	94.1770
Insulin-like growth factor II, human, recombinant. Lyophilized. 5000 units/ampoule.	1st Reference Reagent, 1999	50th Report (in press)	Recombinant growth factor	NIBSC	96/538	99.1898
Interferon alpha, human, leukocyte N3. Lyophilized. 60 000 IU/ampoule.	1st International Standard, 1999	50th Report (in press)	Derived from human leukocytes	NIBSC	95/574	99.1911

Preparation	Standard	WHO TRS ECBS Report	Material	Held at[a]	Code	WHO/BS Documents
Interferon alpha 1/8, human. Lyophilized. 27 000 IU/ampoule.	1st International Standard, 1999	50th Report (in press)	Recombinant cytokine	NIBSC	95/572	99.1911
Interferon alpha 2c, human, recombinant. Lyophilized. 40 000 IU/ampoule.	1st International Standard, 1999	50th Report (in press)	Recombinant cytokine	NIBSC	95/580	99.1911
Interferon alpha consensus, human, recombinant. Lyophilized. 100 000 IU/ampoule.	1st International Standard, 1999	50th Report (in press)	Recombinant cytokine	NIBSC	94/786	99.1911
Interferon alpha, human, leukocyte (HuIFN-a(Le)). Lyophilized. 11 000 IU/ampoule.	2nd International Standard, 1999	50th Report (in press)	Derived from human leukocytes	NIBSC	94/784	99.1911
Interferon alpha, human, lymphoblastoid N1 (HuIFN-alpha(Ly)). Lyophilized. 38 000 IU/ampoule.	2nd International Standard, 1999	50th Report (in press)	Derived from human lymphoblastoid cells	NIBSC	95/568	99.1911
Interferon alpha-1 (alpha-D), human, recombinant, (rHuIFN-alpha1(alphaD)). Lyophilized. 8000 IU/ampoule.	1st International Standard, 1987	No. 771, 38th Report	Recombinant cytokine	NIBSC	83/514	87.1552
Interferon alpha 2a, human, recombinant (rHuIFN-alpha2 (alpha-A)). Lyophilized. 63 000 IU/ampoule.	2nd International Standard, 1999	50th Report (in press)	Recombinant cytokine	NIBSC	95/650	99.1911
Interferon alpha 2b, human, recombinant, (rHuIFN-alpha2 (alpha-2b)). Lyophilized. 70 000 IU/ampoule.	2nd International Standard, 1999	50th Report (in press)	Recombinant cytokine	NIBSC	95/566	99.1911
Interferon beta ser-17, human, recombinant (rHuIFN-beta(ser17)). Lyophilized. 6000 IU/ampoule.	1st International Reference Reagent, 1987	No. 771, 38th Report	Recombinant cytokine	NIAID	Gxb02-901-535	87.1552
Interferon beta, human, fibroblast (HuIFN-beta). Lyophilized. 15000 IU/ampoule.	2nd International Standard, 1987	No. 771, 38th Report	Derived from human fibroblasts	NIAID	Gb23-902-531	87.1552
Interferon gamma, human, recombinant. Lyophilized. 80 000 IU/ampoule.	1st International Standard, 1994	No. 858, 45th Report	Recombinant cytokine	NIAID	Gxg01-902-535	94.1782
Interferon omega, human, recombinant. Lyophilized. 20 000 IU/ampoule.	1st International Standard, 1999	50th Report (in press)	Recombinant cytokine	NIBSC	94/754	99.1911

Preparation	Standard	WHO TRS ECBS Report	Material	Held at[a]	Code	WHO/BS Documents
Interferon, chick. Lyophilized. 80 IU/ampoule.	1st International Reference Preparation, 1978	No. 638, 30th Report	Chick embryo cytokine	NIBSC	67/18	78.1225
Interferon, murine, alpha (MuIFN-alpha). Lyophilized. 16 000 IU/ampoule.	2nd International Standard, 1987	No. 771, 38th Report	Murine cytokine	NIAID	Ga02-901-511	87.1552
Interferon, murine, beta (MuIFN-beta). Lyophilized. 15 000 IU/ampoule.	2nd International Standard, 1987	No. 771, 38th Report	Recombinant cytokine	NIAID	Gb02-902-511	87.1552
Interferon, murine, gamma (MuIFN-gamma). Lyophilized. 1000 IU/ampoule.	1st International Reference Reagent, 1987	No. 771, 38th Report	Recombinant cytokine	NIAID	Gg02-901-533	87.1552
Interferon, rabbit. Lyophilized. 10 000 IU/ampoule.	1st International Standard, 1978	No. 638, 30th Report	Recombinant cytokine	NIAID	G019-902-528	78.1225
Interleukin-1 alpha, human. Lyophilized. 117 000 IU/ampoule.	1st International Standard, 1989	No. 800, 40th Report	Recombinant cytokine	NIBSC	86/632	—
Interleukin-1 beta, human. Lyophilized. 100 000 IU/ampoule.	1st International Standard, 1989	No. 800, 40th Report	Recombinant cytokine	NIBSC	86/680	—
Interleukin-2, human. Lyophilized. 100 IU/ampoule.	1st International Standard, 1987	No. 771, 38th Report	Derived from T-cell line (Jurkat)	NIBSC	86/ 504	87.1559
Interleukin-3, human. Lyophilized. 1700 IU/ampoule.	1st International Standard, 1994	No. 848, 44th Report	Recombinant cytokine	NIBSC	91/510	94.1788
Interleukin-4, human. Lyophilized. 1000 IU/ampoule.	1st International Standard, 1994	No. 858, 45th Report	Recombinant cytokine	NIBSC	88/656	94.1788
Interleukin-5, human. Lyophilized. 5000 units/ampoule.	1st Reference Reagent, 1996	No. 878, 47th Report	Recombinant cytokine	NIBSC	90/586	96.1849
Interleukin-6, human. Lyophilized. 100 000 IU/ampoule.	1st International Standard, 1992	No. 840, 43rd Report	Recombinant cytokine	NIBSC	89/548	92.1713
Interleukin-7, human. Lyophilized. 100 000 units/ampoule.	1st Reference Reagent, 1996	No. 878, 47th Report	Recombinant cytokine	NIBSC	90/530	96.1849

Preparation	Standard	WHO TRS ECBS Report	Material	Held at[a]	Code	WHO/BS Documents
Interleukin-8, human. Lyophilized. 1000 IU/ampoule.	1st International Standard, 1995	No. 872, 46th Report	Recombinant cytokine	NIBSC	89/520	95.1820
Interleukin-9, human. Lyophilized. 1000 units/ampoule.	1st Reference Reagent, 1996	No. 878, 47th Report	Recombinant cytokine	NIBSC	91/678	96.1849
Interleukin-10, human. Lyophilized. 5000 units/ampoule.	1st Reference Reagent, 1997	No. 889, 46th Report	Recombinant cytokine	NIBSC	93/722	97.1868
Interleukin-11, human. Lyophilized. 5000 units/ampoule.	1st Reference Reagent, 1996	No. 878, 47th Report	Recombinant cytokine	NIBSC	92/788	96.1849
Interleukin-12, human. Lyophilized. 10 000 units/ampoule.	1st Reference Reagent, 1996	No. 878, 47th Report	Recombinant cytokine	NIBSC	95/544	96.1849
Interleukin-13, human. Lyophilized. 1000 units/ampoule.	1st Reference Reagent, 1996	No. 878, 47th Report	Recombinant cytokine	NIBSC	94/622	96.1849
Interleukin-15, human. Lyophilized. 10 000 units/ampoule.	1st Reference Reagent, 1996	No. 878, 47th Report	Recombinant cytokine	NIBSC	95/554	96.1849
Islet cell antibodies. Lyophilized. 20 units of islet cell antibodies, 100 units of anti-GAD65 and 100 units of anti-IA-2 per ampoule.	1st Reference Reagent, 1999	50th Report (in press)	Human serum	NIBSC	97/550	99.1896
Kanamycin. Lyophilized. 10 345 IU/ampoule.	1st International Standard, 1986	No. 760, 37th Report	Antibiotic	NIBSC	83/521	86.1515
Kininogenase, porcine, pancreatic. Lyophilized. 22.5 IU/ampoule.	1st International Standard, 1982	No. 687, 33rd Report	Enzyme	NIBSC	78/543	82.1367
Leptin, human. Lyophilized. 4000 IU/ampoule.	1st International Standard, 1999	50th Report (in press)	Recombinant/ *Escherichia coli*	NIBSC	97/564	99.1900
Leptin, mouse. Lyophilized. 4000 IU/ampoule.	1st International Standard, 1999	50th Report (in press)	Recombinant/ *Escherichia coli*	NIBSC	97/626	99.1900
Leukaemia inhibitory factor. Lyophilized. 10 000 units/ampoule.	1st Reference Reagent, 1996	No. 878, 47th Report	Recombinant cytokine	NIBSC	93/562	96.1850
Luteinizing hormone, bovine, for immunoassay. Lyophilized. 0.025 IU/ampoule.	1st International Standard, 1985	No. 745, 36th Report	Pituitary hormone	NIBSC	98/566	85.1474

Preparation	Standard	WHO TRS ECBS Report	Material	Held at[a]	Code	WHO/BS Documents
Luteinizing hormone, human, pituitary. Lyophilized. 35 IU/ampoule.	2nd International Standard, 1988	No. 786, 39th Report	Pituitary hormone	NIBSC	80/552	88.1604
Luteinizing hormone, human, pituitary, alpha subunit. Lyophilized. 10 IU/ampoule.	1st International Standard, 1984	No. 725, 35th Report	Pituitary hormone	NIBSC	78/554	84.1443
Luteinizing hormone, human, pituitary, beta subunit. Lyophilized. 10 IU/ampoule.	1st International Standard, 1984	No. 725, 35th Report	Pituitary hormone	NIBSC	78/556	84.1443
Lymecycline. Lyophilized. 948 IU/mg. Approximately 100 mg.	2nd International Reference Preparation, 1971	No. 486, 24th Report	Antibiotic	NIBSC	68/49	71.1048
Lysine vasopressin. Lyophilized. 7.7 IU/ampoule.	1st International Standard, 1978	No. 638, 30th Report	Synthetic hormone	NIBSC	77/512	78.1230
Macrophage colony stimulating factor, human, recombinant. Lyophilized. 60 000 IU/ampoule.	1st International Standard, 1992	No. 840, 43rd Report	Recombinant growth factor	NIBSC	89/512	92.1712
MAPREC analysis of poliovirus type 3 (Sabin). Lyophilized. 0.9% 472-C nucleotide/vial.	1st International Standard, 1996	No. 878, 47th Report	Chemically synthesized DNA	NIBSC	95/542	96.1841
MAPREC analysis of poliovirus type 3 (Sabin), high virus reference. Lyophilized. 1.1% 472-C nucleotide/vial.	1st Reference Reagent, 1997	No. 889, 48th Report	Chemically synthesized DNA	NIBSC	96/578	97.1865
MAPREC analysis of poliovirus type 3 (Sabin), low virus reference. Lyophilized. 0.7% 472-C nucleotide/vial.	1st Reference Reagent, 1997	No. 889, 48th Report	Chemically synthesized DNA	NIBSC	96/572	97.1865
Measles vaccine (live). Lyophilized. 4.4 $\log_{10}$ (20 000) infectious units/vial.	2nd International Reference Reagent, 1994	No. 858, 45th Report	Attenuated measles virus	NIBSC	92/648	94.1771
Methacycline. Lyophilized. 924 IU/mg. Approximately 50 mg of methacycline hydrochloride.	1st International Reference Preparation, 1969	No. 444, 22nd Report	Antibiotic	NIBSC	68/319	69.994
Mumps vaccine (live). Lyophilized. 4.6 $\log_{10}$ (40 000) infectious units/vial.	1st International Reference Reagent, 1994	No. 858, 45th Report	Attenuated mumps virus	NIBSC	90/534	94.1772

Preparation	Standard	WHO TRS ECBS Report	Material	Held at[a]	Code	WHO/BS Documents
***Naja* (*Naja* & *Hemachatus* species) antivenin, equine**. Lyophilized. 300 IU/ampoule.	1st International Standard, 1964	No. 293, 17th Report	Horse serum	NIBSC	NAJA	64.708
Neomycin. Lyophilized. 775 IU/mg. Ampoules containing approximately 50 mg of neomycin sulfate.	2nd International Reference Preparation, 1975	No. 594, 27th Report	Antibiotic	NIBSC	72/406	75.1097
Neomycin B. Lyophilized. 670 IU/mg. Ampoules containing approximately 25 mg of neomycin B sulfate.	1st International Reference Preparation, 1975	No. 594, 27th Report	Antibiotic	NIBSC	68/41	75.1098
Nerve growth factor. Lyophilized. 10 000 units/ampoule.	1st Reference Reagent, 1996	No. 878, 47th Report	Recombinant growth factor	NIBSC	93/556	96.1836
Netilmicin. Lyophilized. 4810 IU/ampoule.	1st International Standard, 1989	No. 800, 40th Report	Antibiotic	NIBSC	83/577	89.1628
Newcastle (Hitchner B1 strain) disease vaccine, live. Lyophilized. No unitage assigned.	1st International Reference Preparation, 1967	No. 384, 20th Report	Attenuated Newcastle virus	NIBSC	NV	67.882
Newcastle disease vaccine (inactivated). Lyophilized. 525 IU/ampoule.	1st International Standard, 1963	No. 274, 16th Report	Antigen	NIBSC	NVIA	63.626, 64.671
Novobiocin. Lyophilized. 970 IU/mg. Approximately 100 mg of novobiocin acid.	1st International Standard, 1965	No. 329, 18th Report	Antibiotic	NIBSC	62/9	65.766
Nystatin. Lyophilized. 4855 IU/mg. Approximately 100 mg.	2nd International Standard, 1982	No. 687, 33rd Report	Antibiotic	NIBSC	80/508	82.1350
Oncostatin M. Lyophilized. 25 000 units/ampoule.	1st Reference Reagent, 1996	No. 878, 47th Report	Recombinant cytokine	NIBSC	93/564	96.1851
Opacity. 10 IU of opacity.	5th International Standard, 1975	No. 594, 27th Report	Miscellaneous	NIBSC	76/522	75.1119
Oxytocin. Lyophilized. 12.5 IU/ampoule.	4th International Standard, 1978	No. 638, 30th Report	Synthetic oxytocin peptide	NIBSC	76/575	78.1227
Parathyroid hormone, bovine. Lyophilized. 39 IU/ampoule.	1st International Standard, 1985	No. 745, 36th Report	Parathyroid peptide hormone	NIBSC	82/632	85.1490

Preparation	Standard	WHO TRS ECBS Report	Material	Held at[a]	Code	WHO/BS Documents
Parathyroid hormone, human, for immunoassay. Lyophilized. 0.1 IU/ampoule.	1st International Reference Preparation, 1981	No. 673, 32nd Report	Parathyroid peptide hormone	NIBSC	79/500	81.1315
Paromomycin. Approximately 75 mg of paromomycin sulfate.	1st International Reference Preparation, 1966	No. 361, 19th Report	Antibiotic	NIBSC	65/61	66.822
Pertussis vaccine. Lyophilized. 46 IU/ampoule.	3rd International Standard, 1998	No. 897, 49th Report	Inactivated *Bordetella pertussis*	NIBSC	66/303	98.1880
Placental lactogen, human, for immunoassay. Lyophilized. 0.000 850 IU/ampoule.	1st International Reference Preparation, 1977	No. 626, 29th Report	Purified placental protein	NIBSC	73/545	77.1141
Plasmin, human. Lyophilized. 5.3 IU/ampoule.	3rd International Standard, 1998	No. 897, 49th Report	Human plasma protein	NIBSC	97/536	98.1887
Plasminogen-activator inhibitor 1 (PAI-1), human. Lyophilized. Tissue-plasminogen-activator: 27.5 IU of neutralizing activity per ampoule. Urinary-plasminogen-activator: 7 IU of neutralizing activity per ampoule.	1st International Standard, 1995	No. 872, 46th Report	Recombinant protein (Chinese hamster ovary cells), spiked in plasma	NIBSC	92/654	95.1805
Platelet-derived growth factor-BB isoform. Lyophilized. 3000 IU/ampoule.	1st International Standard, 1997	No. 889, 48th Report	Recombinant growth factor	NIBSC	94/728	97.1864
Platelet factor 4. Lyophilized. 400 IU/ampoule.	1st International Standard, 1984	No. 725, 35th Report	Human platelet protein	NIBSC	83/505	84.1455
Poliomyelitis vaccine (inactivated). Lyophilized. Type 1 antigen: 430 D-antigen units/ml; type 2 antigen: 95 D-antigen units/ml; type 3 antigen: 285 D-antigen units/ml.	1st International Reference Reagent, 1994	No. 858, 45th Report	Inactivated polioviruses	NIBSC	91/574	94.1777, 94.1777 Corr. 1, 94.1778
Poliovirus (Sabin), live attenuated. Lyophilized. Type 1: 6.6 $\log_{10}$ $CCID_{50}$/ml ; type 2: 5.6 $\log_{10}$ $CCID_{50}$/ml ; type 3: 6.2 $\log_{10}$ $CCID_{50}$/ml; total virus content: 6.8 $\log_{10}$ $CCID_{50}$/ml.	1st International Reference Reagent, 1995	No. 872, 46th Report	Attenuated polioviruses	NIBSC	85/659	95.1821

Preparation	Standard	WHO TRS ECBS Report	Material	Held at[a]	Code	WHO/BS Documents
Polymyxin B. 8403 IU/mg. Approximately 75 mg of purified polymyxin B sulfate.	2nd International Standard, 1969	No. 444, 22nd Report	Antibiotic	NIBSC	67/301	69.990
Prekallikrein activator. Lyophilized. 85 IU/ampoule.	1st International Standard, 1984	No. 725, 35th Report	Purified from human plasma	NIBSC	82/530	84.1454
Proinsulin, bovine, for immunoassay. Lyophilized. 25 µg/ampoule.	1st International Standard, 1986	No. 760, 37th Report	Pancreatic enzyme	NIBSC	84/514	86.1534
Proinsulin, human. Lyophilized. 6 µg/ampoule.	1st International Reference Reagent, 1986	No. 760, 37th Report	Recombinant proinsulin	NIBSC	84/611	86.1537
Proinsulin, porcine, for immunoassay. Lyophilized. 20 µg/ampoule.	1st International Standard, 1986	No. 760, 37th Report	Pancreatic enzyme	NIBSC	84/528	86.1534
Prolactin, human. Lyophilized. 0.053 IU/ampoule.	3rd International Standard, 1988	No. 786, 39th Report	Pituitary hormone	NIBSC	84/500	86.1520, 86.1520 Add. 1, 88.1596
Prolactin, ovine. Lyophilized. 22 IU/mg. Approximately 10 mg/ampoule.	2nd International Standard, 1962	No. 259, 15th Report	Pituitary hormone	NIBSC	57/8	62.577
Prostate-specific antigen. Lyophilized. 1 µg total PSA per vial.	1st Reference Reagent, 1999	50th Report (in press)	Purified plasma protein	NIBSC	96/668	99.1902
Prostate-specific antigen (90:10). Lyophilized. 1 µg total PSA per vial.	1st Reference Reagent, 1999	50th Report (in press)	Purified plasma protein	NIBSC	96/700	99.1902
Protamine. Lyophilized. Approximately 60 mg.	1st International Reference Preparation, 1954	No. 96, 8th Report	Salmon protamine	NIBSC	54/5	54.261
Protein C, plasma, human. Lyophilized. 0.82 IU/ampoule.	1st International Standard, 1987	No. 771, 38th Report	Human plasma	NIBSC	86/622	87.1561
Protein S, plasma, human. Lyophilized. 0.90 IU/ampoule.	1st International Standard, 1995	No. 872, 46th Report	Human plasma	NIBSC	93/590	95.1804

Preparation	Standard	WHO TRS ECBS Report	Material	Held at[a]	Code	WHO/BS Documents
Pyrogen. Lyophilized. Approximately 2 mg/vial.	1st International Reference Preparation, 1957	No.147, 11th Report	"O" somatic antigen of *Shigella dysenteriae*	NIBSC	57/7	57.400, 82.1372
Rabies vaccine. Lyophilized. 16 IU/ampoule (10 IU of PM-glycoprotein and 135 IU of PM-ribonucleoprotein per ampoule, in addition to the 16 IU of rabies vaccine).	5th International Standard, 1991	No. 822, 42nd Report	Rabies inactivated virus	NIBSC	RAV	91.1654, 91.1659, 91.1661
Renin, human. Lyophilized. 0.1 IU/ampoule.	1st International Reference Preparation, 1974	No. 565, 26th Report	Derived from human kidney extract	NIBSC	68/356	74.1089
Rheumatoid arthritis serum. Lyophilized. 100 IU/ampoule.	1st International Reference Preparation, 1973	No. 530, 25th Report	Human serum	CLB	W1066	73.1067
Rifamycin SV. Lyophilized. 887 IU/mg. Approximately 100 mg of sodium rifamycin SV.	1st International Reference Preparation, 1967	No. 384, 20th Report	Antibiotic	NIBSC	66/231	67.885
Rubella vaccine (live). Lyophilized. 3.9 $\log_{10}$ (8000) infectious units/vial.	1st International Reference Reagent, 1994	No. 858, 45th Report	Attenuated rubella virus	NIBSC	91/688	94.1773
Scarlet fever streptococcus antitoxin, equine. Lyophilized. 10 000 IU/ampoule.	1st International Standard, 1952	No. 68, 6th Report	Horse serum	NIBSC	SC	52.150, 53.225
Serum amyloid A protein. Lyophilized. 0.15 IU/ampoule.	1st International Standard, 1997	No. 889, 48th Report	Human serum	NIBSC	92/680	97.1860
Sex hormone binding globulin (SHBG). Lyophilized. 107 IU/ampoule.	1st International Standard, 1998	No. 897, 49th Report	Human serum	NIBSC	95/560	98/1884
Short ragweed pollen extract (*Ambrosia elatior*). Lyophilized. 100 000 IU/ampoule.	1st International Standard, 1983	No. 700, 34th Report	Defatted short ragweed pollen	NIBSC	84/581	83.1412
Sisomicin. Lyophilized. 35 200 IU/ampoule.	1st International Standard, 1984	No. 725, 35th Report	Antibiotic	NIBSC	80/543	84.1434

Preparation	Standard	WHO TRS ECBS Report	Material	Held at[a]	Code	WHO/BS Documents
Smallpox vaccine. Lyophilized. 14 mg/ampoule. No unitage assigned.	1st International Reference Preparation, 1962	No. 259, 15th Report	Antigen	NIBSC	SMV	61.536, 62.546
Somatropin (rDNA-derived human growth hormone). Lyophilized. 2.0 mg protein/ampoule, 3.0 IU/mg.	1st International Standard, 1994	No. 858, 45th Report	Recombinant/ *Escherichia coli*	NIBSC	88/624	93.1735, 94.1790
Spiramycin. Lyophilized. 3200 IU/mg. Approximately 50 mg.	1st International Reference Preparation, 1964	No. 293, 17th Report	Antibiotic	NIBSC	62/8	64.692
***Staphylococcus*-alpha antitoxin, equine**. Lyophilized. 220 IU/ampoule.	3rd International Standard, 1982	No. 687, 33rd Report	Horse serum	NIBSC	STA	82.1345
Stem cell factor. Lyophilized. 1000 units/ampoule.	1st Reference Reagent, 1997	No. 889, 48th Report	Recombinant cytokine	NIBSC	91/682	97.1859
Streptokinase. Lyophilized. 700 IU/ampoule.	2nd International Standard, 1989	No. 800, 40th Report	*Streptococcus haemolyticus*	NIBSC	88/826	89.1623
Streptomycin. Lyophilized. 78 500 IU/ampoule.	3rd International Standard, 1980	No. 658, 31st Report	Antibiotic	NIBSC	76/539	80.1273
Swine erysipelas serum (anti-N). Lyophilized. 628 IU/ampoule.	1st International Standard, 1954	No. 96, 8th Report	Horse serum	NIBSC	SES	54.297, 54.300
Teicoplanin. Lyophilized. 51 550 IU/ampoule.	1st International Standard, 1990	No. 814, 41st Report	Antibiotic	NIBSC	90/704	90.1642
Tetanus toxoid, adsorbed. Lyophilized. 340 IU/ampoule.	2nd International Standard, 1981	No. 673, 32nd Report	Adsorbed toxoid	NIBSC	TEXA	81.1311, 83.1395
Tetanus toxoid, for flocculation tests. Lyophilized. 1000 Lf/ampoule.	1st International Reference Reagent, 1988	No. 786, 39th Report	Toxoid	NIBSC	TEFT	88.1590

Preparation	Standard	WHO TRS ECBS Report	Material	Held at[a]	Code	WHO/BS Documents
Tetracosactrin. Lyophilized. 490 IU/ampoule.	1st International Reference Preparation, 1981	No. 673, 32nd Report	Synthetic peptide	NIBSC	80/590	81.1313
Thrombin, alpha, human. Lyophilized. 100 IU/ampoule.	1st International Standard, 1991	No. 822, 42nd Report	Purified plasma protein	NIBSC	89/588	91.1669
Thromboplastin, bovine, combined. Lyophilized. International sensitivity index: 1.0.	2nd International Reference Preparation, 1983	No. 700, 34th Report	Derived from brain	CLB	OBT/79	83.1393
Thromboplastin, human, recombinant, plain. Lyophilized. International sensitivity index: 0.940.	3rd International Reference Preparation, 1996	No. 878, 47th Report	Recombinant protein	CLB	rTF/95	96.1846 Rev. 1
Thromboplastin, rabbit, plain. Lyophilized. International sensitivity index: 1.0.	3rd International Reference Reagent, 1995	No. 872, 46th Report	Derived from brain	CLB	RBT/90	95.1796
Thyroid-stimulating antibody. Lyophilized. 0.1 IU/ampoule.	1st International Reference Preparation, 1995	No. 872, 46th Report	Human serum	NIBSC	90/672	95.1818
Thyroid-stimulating hormone, human, for immunoassay. Lyophilized. 0.037 IU/ampoule.	2nd International Reference Preparation, 1983	No. 700, 34th Report	Pituitary hormone	NIBSC	80/558	83.1402
Thyroid-stimulating hormone, human, recombinant. Lyophilized. 0.0067 units/ampoule.	1st Reference Reagent, 1996	No. 878, 47th Report	Recombinant hormone (Chinese hamster ovary cells)	NIBSC	94/674	96.1843
Thyrotrophin, bovine. Lyophilized. 0.074 IU/mg.	1st International Standard, 1955	No. 108, 9th Report	Pituitary hormone	NIBSC	53/11	55.309
Thyroxine-binding globulin. Lyophilized. 30 IU/ampoule.	1st International Standard, 1991	No. 822, 42nd Report	Derived from human serum	NIBSC	88/638	91.1671
Timothy (*Phleum pratense*) pollen extract. Lyophilized. 100 000 IU/ampoule.	1st International Standard, 1983	No. 700, 34th Report	*Phleum pratense* allergenic extract	NIBSC	82/520	83.1411

Preparation	Standard	WHO TRS ECBS Report	Material	Held at[a]	Code	WHO/BS Documents
Tissue plasminogen activator (t-PA), human, recombinant. Lyophilized. 10 000 IU/ampoule.	1st International Standard, 1999	50th Report (in press)	r-tPA (Chinese hamster ovary cells)	NIBSC	98/714	99.1913
Tobramycin. Lyophilized. 9800 IU/ampoule.	2nd International Standard, 1985	No. 745, 36th Report	Antibiotic	NIBSC	82/510	85.1504
Tuberculin, old. Liquid. 90 000 IU/ml.	3rd International Standard, 1965	No. 329, 18th Report	Derived from *Mycobacterium tuberculosis* culture	NIBSC	TU	65.779
Tuberculin, purified protein derivative (PPD), avian. Lyophilized. 500 000 IU/ampoule.	1st International Standard, 1954	No. 96, 8th Report	Derived from *Mycobacterium avium* culture	NIBSC	PPDA	53.227, 54.293
Tuberculin, purified protein derivative (PPD), bovine. Lyophilized. 58 500 IU/ampoule.	1st International Standard, 1986	No. 760, 37th Report	Derived from *Mycobacterium bovis* culture	NIBSC	PPD BOV	86.1518
Tuberculin, purified protein derivative (PPD), mammalian. Lyophilized. 5000 IU/ampoule.	1st International Standard, 1951	No. 56, 5th Report	Derived from *Mycobacterium tuberculosis* culture	NIBSC	PPDT	51.127, 83.1408
Tumour necrosis factor, alpha, human. Lyophilized. 40 000 IU/ampoule.	1st International Standard, 1991	No. 822, 42nd Report	Recombinant cytokine	NIBSC	87/650	91.1681
Tumour necrosis factor, beta, human. Lyophilized. 150 000 units/ampoule.	1st Reference Reagent, 1996	No. 878, 47th Report	Recombinant cytokine	NIBSC	87/640	96.1852
Typhoid vaccine (acetone-inactivated). Lyophilized. No unitage assigned.	1st International Reference Preparation, 1960	No. 222, 14th Report	Inactivated *Salmonella typhi*	NIBSC	TYVK	60.515, 68.906
Typhoid vaccine (heat-phenol-inactivated). Lyophilized. No unitage assigned.	1st International Reference Preparation, 1960	No. 222, 14th Report	Inactivated *Salmonella typhi*	NIBSC	TYVL	60.515, 68.906
Urokinase, high molecular weight. Lyophilized. 4300 IU/ampoule.	1st International Standard, 1989	No. 800, 40th Report	Derived from two-chain urine	NIBSC	87/594	89.1622

Preparation	Standard	WHO TRS ECBS Report	Material	Held at[a]	Code	WHO/BS Documents
Vancomycin. Lyophilized. 1007 IU/mg. Approximately 50 mg of vancomycin sulfate.	1st International Standard, 1963	No. 274, 16th Report	Antibiotic	NIBSC	59/20	63.648
Vitamin B_{12}, in human serum. Lyophilized. 320 pg/ampoule.	1st International Reference Reagent, 1992	No. 840, 43rd Report	Human serum	NIBSC	81/563	92.1703
Whole blood folate. Lyophilized. 13 ng/ampoule.	1st International Standard, 1996	No. 878, 47th Report	Haemolysed blood	NIBSC	95/528	95.1806, 96.1839

WHO Technical Report Series, No. 897, 2000

Annex 5

Biological substances: International Standards and Reference Reagents

At its meeting in October 1998, the WHO Expert Committee on Biological Standardization made a number of changes to the previously published list,[1] which are set out below. A list of current International Biological Reference Preparations, incorporating these changes, is published as a separate annex (Annex 4) to this report.

Additions

Antibodies

Anti-hepatitis A immunogobulin, human	49 IU/ampoule	Second International Standard 1998
Clostridium perfringens beta antitoxin, equine	4770 IU/ampoule	Second International Standard 1998

Antigens

Pertussis vaccine	46 IU/ampoule	Third International Standard 1998

Blood products

Blood coagulation factor VII, concentrate, human	6.3 IU/ampoule	First International Standard 1998
Blood coagulation factor VIII, concentrate, human	8.5 IU/ampoule	Sixth International Standard 1998
Blood coagulation factor VIII and von Willebrand factor, plasma, human	0.57 IU/ampoule for factor VIII clotting activity 0.89 IU/ampoule for factor VIII antigen 0.79 IU/ampoule for von Willebrand antigen 0.73 IU/ampoule for von Willebrand restocetin cofactor activity	Fourth International Standard 1998
Haemoglobincyanide	49.79 μmol/l	Sixth International Standard 1998
Heparin, unfractionated	2031 IU/ampoule	Fifth International Standard 1998
Plasmin, human	5.3 IU/ampoule	Third International Standard 1998

[1] *Biological substances: International Standards and Reference Reagents 1990.* Geneva, World Health Organization, 1991.

Cytokines

Activin A, human, recombinant	5 units/ampoule	First Reference Reagent 1998

Endocrinological substances

Sex hormone binding globulin	107 IU/ampoule	First International Standard 1998

Discontinuations

Antibiotics

Demeclocycline	First International Reference Preparation 1962
Doxycycline	First International Reference Preparation 1973
Hygromycin B	First International Reference Preparation 1966
Minocycline	First International Reference Preparation 1975
Oxytetracycline	First International Reference Preparation 1966
Tetracycline	First International Reference Preparation 1970

Antibodies

Histoplasmin antiserum, rabbit, for H and M immunodiffusion test	First Reference Reagent 1981
Mycoplasma pneumoniae antiserum, equine	First Reference Reagent 1981
Parainfluenza virus, antiserum, equine	First Reference Reagent 1981

Miscellaneous

Histoplasmin, for H and M immunodiffusion test	First Reference Reagent 1981
Hyaluronindase	First International Standard 1955

WHO Technical Report Series, No. 897, 2000

Annex 6

Corrigendum: International Standard for Insulin-like Growth Factor 1

The potency assigned to the International Standard for Insulin-like Growth Factor 1 in WHO Technical Report Series, No. 858, 1995, should be amended as follows:

Page 17, line 40: *replace...* 150 000 International Units... *with...* 150 International Units...

Page 94, line 28: *replace...* 150 000 IU/ampoule... *with...* 150 IU/ampoule...

World Health Organization Technical Report Series

Recent reports: No.		Sw.fr.*
851	(1995) **Evaluation of certain veterinary drug residues in food** Forty-second report of the Joint FAO/WHO Expert Commitee on Food Additives (50 pages)	10.–
852	(1995) **Onchocerciasis and its control** Report of a WHO Expert Committee on Onchocerciasis Control (111 pages)	15.–
853	(1995) **Epidemiology and prevention of cardiovascular diseases in elderly people** Report of a WHO Study Group (72 pages)	14.–
854	(1995) **Physical status: the use and interpretation of anthropometry** Report of a WHO Expert Committe (462 pages)	71.–
855	(1995) **Evaluation of certain veterinary drug residues in food** Forty-third report of the Joint FAO/WHO Expert Committee on Food Additives (65 pages)	
856	(1995) **WHO Expert Committee on Drug Dependence** Twenty-ninth report (21 pages)	6.–
857	(1995) **Vector control for malaria and other mosquito-borne diseases** Report of a WHO Study Group (97 pages)	15.–
858	(1995) **WHO Expert Committee on Biological Standardization** Forty-fifth report (108 pages)	17.–
859	(1995) **Evaluation of certain food additives and contaminants** Forty-fourth report of the Joint FAO/WHO Expert Committee on Food Additives (62 pages)	11.–
860	(1996) **Nursing practice** Report of a WHO Expert Committee (37 pages)	12.–
861	(1996) **Integration of health care delivery** Report of a WHO Study Group (74 pages)	14.–
862	(1996) **Hypertension control** Report of a WHO Expert Committee (89 pages)	16.–
863	(1996) **WHO Expert Committee on Specifications for Pharmaceutical Preparations** Thirty-fourth report (200 pages)	35.–
864	(1996) **Evaluation of certain veterinary drug residues in food** Forty-fifth report of the Joint FAO/WHO Expert Committee on Food Additives (66 pages)	12.–
865	(1996) **Control of hereditary diseases** Report of a WHO Scientific Group (89 pages)	16.–
866	(1996) **Research on the menopause in the 1990s** Report of a WHO Scientific Group (114 pages)	20.–
867	(1997) **The use of essential drugs** Seventh report of the WHO Expert Committee (80 pages)	15.–
868	(1997) **Evaluation of certain food additives and contaminants** Forty-sixth report of the Joint FAO/WHO Expert Committee on Food Additives (77 pages)	14.–
869	(1997) **Improving the performance of health centres in district health systems** Report of a WHO Study Group (70 pages)	14.–
870	(1997) **Promoting health through schools** Report of a WHO Expert Committee on Comprehensive School Health Education and Promotion (99 pages)	20.–

* Prices in developing countries are 70% of those listed here.

871 (1997) **Medical methods for termination of pregnancy**
Report of a WHO Scientific Group (117 pages) 23.–

872 (1998) **WHO Expert Committee on Biological Standardization**
Forty-sixth report (97 pages) 20.–

873 (1998) **WHO Expert Committee on Drug Dependence**
Thirtieth report (56 pages) 14.–

874 (1998) **WHO Expert Committee on Leprosy**
Seventh report (49 pages) 14.–

875 (1998) **Training in diagnostic ultrasound: essentials, principles and standards**
Report of a WHO Study Group (52 pages) 14.–

876 (1998) **Evaluation of certain veterinary drug residues in food**
Forty-seventh report of the Joint FAO/WHO Expert Committee on Food Additives (91 pages) 19.–

877 (1998) **Cardiovascular disease and steroid hormone contraception**
Report of a WHO Scientific Group (96 pages) 20.–

878 (1998) **WHO Expert Committee on Biological Standardization**
Forty-seventh report (107 pages)

879 (1998) **Evaluation of certain veterinary drug residues in food**
Forty-eighth report of the Joint FAO/WHO Expert Committee on Food Additives (80 pages) 16.–

880 (1998) **Preparation and use of food-based dietary guidelines**
Report of a Joint FAO/WHO consultation (114 pages) 23.–

881 (1998) **Control and surveillance of African trypanosomiasis**
Report of a WHO Expert Committee (119 pages) 23.–

882 (1998) **The use of essential drugs**
Eighth report of the WHO Expert Committee (83 pages) 19.–

883 (1999) **Food safety issues associated with products from aquaculture**
Report of a Joint FAO/NACA/WHO Study Group (62 pages) 14.–

884 (1999) **Evaluation of certain food additives and contaminants**
Report of a Joint FAO/WHO Expert Committee on Food Additives (104 pages) 20.–

885 (1999) **WHO Expert Committee on Specifications for Pharmaceutical Preparations**
Thirty-fifth report (162 pages) 35.–

886 (1999) **Programming for adolescent health and development**
Report of a WHO/UNFPA/UNICEF Study Group (266 pages) 56.–

887 (1999) **WHO Expert Committee on Drug Dependence**
Thirty-first report (28 pages) 14.–

888 (1999) **Evaluation of certain veterinary drug residues in food**
Fiftieth report of the Joint FAO/WHO Expert Committee on Food Additives (102 pages) 20.–

889 (1999) **WHO Expert Committee on Biological Standardization**
Forty-eighth report (117 pages) 23.–

890 (1999) **High-dose irradiation: wholesomeness of food irradiated with doses above 10 kGy**
Report of a Joint FAO/IAEA/WHO Study Group (203 pages) 42.–

891 (2000) **Evaluation of certain food additives**
Fifty-first report of the Joint FAO/WHO Expert Committee on Food Additives (176 pages) 35.–

892 (2000) **WHO Expert Committee on Malaria**
Twentieth report (76 pages) 14.–

893 (2000) **Evaluation of certain veterinary drug residues in food**
Fifty-second report of the Joint FAO/WHO Expert Committee on Food Additives (109 pages) 20.–

894 (2000) **Obesity: preventing and managing the global epidemic**
Report of a WHO Consultation (264 pages) 56.–

895 (2000) **The use of essential drugs**
Ninth report of the WHO Expert Committee on the Use of Essential Drugs (including the revised Model List of Essential Drugs) (65 pages) 14.–

896 (2000) **Evaluation of certain food additives and contaminants**
Fifty-third report of the Joint FAO/WHO Expert Committee on Food Additives (136 pages) 25.–